SBHMC BOOK
54009000090825

D1345467

The Biostatistics Cookbook

The most user-friendly guide for the bio/medical scientist

Seth Michelson
Roche Bioscience, Palo Alto, CA, USA

and

Timothy Schofield
Merck Research Laboratories, West Point, PA, USA

KLUWER ACADEMIC PUBLISHERS
DORDRECHT / BOSTON / LONDON

Distributors

---

for the United States and Canada: Kluwer Academic Publishers, PO Box 358, Accord Station, Hingham, MA 02018-0358, USA
for all other countries: Kluwer Academic Publishers Group, Distribution Center, PO Box 322, 3300 AH Dordrecht, The Netherlands

A C.I.P. Catalogue record for this book is available from the Library of Congress.

ISBN 0-7923-3884-7 (hard cover)
ISBN 0-7923-4105-8 (limp cover)

Printed on acid-free paper

Kluwer Academic Publishers BV incorporates the publishing programmes of D. Reidel, Martinus Nijhoff, Dr W. Junk and MTP Press.

Printed in Great Britain by Hartnolls Ltd., Bodmin, Cornwall.

Contents

INTRODUCTION

We live in a very uncertain world. Variation surrounds our work. There is noise in our experiments, in our measurements, and in our test subjects. From all these sources of uncertainty and variation, we try to extract a coherent picture of very complex and sometimes dynamic, biological and chemical processes. In fact, one of our major challenges is to separate this signal, the 'real' biology or chemistry, from the noise. The tools developed to do this are called, collectively, biostatistics.

Any tool, even a hammer, can be misused. This could result, at best, in inefficiency, and, at worst, in disaster. With the advent of newer, user-friendly statistical software packages, desk top computing, and point-and-click technologies, it is easier than ever to make mistakes in your analyses. The beauty of having access to so much computing power is that you can now enjoy ultimate flexibility in data processing: that can also be a problem. Ask your computer to produce a particular analysis, report or graphic, and that is exactly what you will get: if you happen to have asked for the wrong thing it will be produced just as quickly, and you will probably never know it was wrong. One aim of this handbook is to help you choose the correct tool for the job at hand, understand its strengths and weaknesses, and to help you recognize when you should seek expert advice.

We describe biostatistics as a collection of tools for very good reasons. They are techniques that have been developed to do a job. Although the mathematical theory behind them can sometimes be rather esoteric and quite complex, our primary concern, as experimental scientists, is on how they may be applied, not on the theory behind them.

We use biostatistics – the entire tool box – to achieve a variety of goals. We can use some of these tools to describe our data in standard, rigorous ways which allow our audience to know exactly what we mean, and do not mean, when we discuss our results. Other tools are used to compare and draw inferences about populations: a word that needs to be taken in its broadest sense. Animals treated with different drugs represent different populations, but so do stones quarried from different sites. Yet another set of tools can be used to derive estimates of model parameters. A dose–response curve is a good example of a model based system from which estimates for parameters such as the ED_{50} or LD_{10} can be derived. These estimation tools can also provide a good insight into how much uncertainty there is in the model, the data, etc. and how much faith should be placed in the results. The main categories we have just described are called *description, inference* and *estimation,* and we will devote one chapter to each.

The point of this book is to make Biostatistics accessible. We want to inflame your intuition. Biostatistics can be intimidating if all you see are mathematical formulae – but if you understand why a particular test is performed and what it means in plain English, then you will know when and how to apply it to your own particular problems. That is our goal!

1 DESCRIPTION

Collections of data are not the same thing as information. This is a rather harsh generalization, but one which holds when examined critically. Data points are measurements; they are random 'snapshots' of random processes. Because we human beings are limited by our technology, our measurements contain errors, and because it is impossible to run an experiment of infinite scope and range, data obtained from a limited sample must be extended to an entire underlying population. Data are, therefore, inherently noisy and incomplete.

Information, on the other hand, depends upon context. Data need to be interpretable within that context. Valid summary and description are required to allow the signal to be separated from the noise and to enable the information obtained to be shared. For example, it makes no sense to separate your subjects into different classes and then ignore these classifications when you summarize your results. There must have been a reason for separating them in the first place: either they received different treatments, they represent different kinds of people, perhaps men and women, or they display some other attribute that makes them unique. In the next chapter we will explore ways of comparing groups. Before we do, however, it is important that you become acquainted with your data – summarize it, display it and extract from it all the information it has to offer. The tools of biostatistics which allow you to summarize, plot and interpret your data are called *descriptive statistics*. In the following sections we will discuss each tool separately, but first we will present a brief overview of the areas to be covered.

The point of data description is to enable communication with your colleague – but what do you want to tell them? Do you really just want to describe the single sample of 10 rats you just received from your animal colony, or do you want to describe the class of subjects known as 'rat' and the effects of a particular treatment upon them? In order to generalize from your sample to the whole population you must be able to associate your observed data with an ideal underlying population that represents all the rats you could have possibly tested. In other words, we need to separate in our own minds the idea of 'population' from the idea of 'sample' so that we can derive a description of the first from the second.

What do we mean by a description? Typically, we want to tell our audience about how our population responds to a stimulus. We would like to say something about the average behavior we observe, whether we mean blood pressure in rats or densities in rocks. The statistician (and the skeptic!) usually also wants to know how your data are distributed around the average. Is one value, or set of values, more likely to occur than any other? We also need to know how much noise is inherent in the experiment.

Suppose you could study simultaneously all the spontaneously hypertensive rats in the world. You might observe some with mean blood pressures below 90mmHg, although the chances of that happening are quite small, maybe even 1 in a million. You would probably see more rats with blood pressures between 90 and 100 mmHg, and more still between 100 and 110 mmHg. If you allocated every hypertensive rat in the world to a group defined by blood pressure, classified in 10 mmHg intervals from 90 to 300 mmHg, you would have a clear picture of your population. That kind of experiment cannot be performed and reported in any reasonable time. You therefore need to say something about rats based upon the data observed in, say, 10 of their representatives. In the next section we will discuss populations, samples and distributions, and tie them together so that the summaries you derive from your sample actually represent the underlying population in a statistically rigorous way.

POPULATIONS, DISTRIBUTIONS AND SAMPLES

Terms you should learn:

Target population
Statistical population
Sample population
Underlying distribution
Sample distribution
Observations

Concepts you should master:

Generalizations from sample to statistic to target
Frequencies, probabilities and events
Random sampling
Bias

The average person uses the word 'population' to mean a collection of individuals living together in a community. To the statistician, though, the word means much more than that. Formally, *a statistical population* is the set of all possible values (called observations) that could be obtained for a given attribute if all the test subjects were measured simultaneously. Less formally, suppose you are interested in a population of hypertensive rats, and suppose you decide to measure one attribute that you think describes your rats, say blood pressure or heart rate. The entire range of all possible blood pressures makes up the statistical population. While the point is a subtle one, it deserves to be made. You want to describe *a target population* (hypertensive rats) by summarizing a set of measures (blood pressure) and generalize from one back to the other. It is the population of blood pressure values which interests the statistician.

Let us consider other examples. Suppose you were measuring the density of igneous rock. Then the statistical population of interest is not all the igneous rocks in the world, but all their densities. The target population you want to describe is 'igneous rock' by summarizing the attribute we call 'density'. Suppose you want to verify the quality of an assay run for you by an outside laboratory. The target population would be all the tests run for you by that laboratory, and the statistical population might be all hemoglobin measurements performed during January.

Care is needed, however. A target population and an attribute do not necessarily have anything to do with each other. For example, in the most absurd case, you could measure the tail lengths of hypertensive rats rather than their blood pressures. One must wonder why, but if you did do something so silly, why would you target hypertensives rather than normotensives? Do you really gain any insight into your target population that you would not have had anyway? What you really want to summarize (and then tell your colleagues about) is blood pressure. Maybe you want to describe new blood pressure lowering medicines, or maybe just the rat population itself. In either case, tail length will probably not suffice since it is not a 'surrogate' for blood pressure. Good statistics cannot help silly science and vice versa!

If we assume that you choose a statistical population that really represents your target, the next step is to build the link between your target and statistical populations, i.e. to define a mathematically descriptive relationship between your subjects and your statistical universe. If we could count the number of subjects in the entire universe that achieves a value between some predefined upper and lower limit, and if we let these intervals cover our entire universe, then we could calculate the frequency of observations within each interval. From that set of frequencies we would know exactly what the most frequently attained values are. The whole set of frequency–value pairs makes up what the statistician calls *the underlying distribution* of the statistical population. Grouping the observations into predefined intervals, counting their frequencies and presenting them graphically results in a plot known as the histogram, which is covered in much greater detail below.

Mathematically, the frequency distribution of the underlying population explicitly defines a probability space. That means that we now know the exact chances of a value drawn from any subject falling within a specified interval. To carry our hypertensive rat example to its most extreme limits, we know that if 23% of all the hypertensive rats in the world registered mean blood pressures between 140 and 150 mmHg, the chances of observing any one rat with a measure in that range is 23/100. The frequency distribution therefore becomes a measure of probability in *an event space* where the events are 'blood pressure between . . .'. This linkage between the underlying frequency distribution and the probability of observing any particular event, e.g. blood pressure between 140 mmHg and 150 mmHg, forms the basis for the inferential statistics presented below.

You have probably heard of terms such as normal or Gaussian distribution, chi-square distribution, F-distribution. These are simply well defined probability distributions which seem to describe the real world fairly well. Each is well established and well characterized. More importantly, each has been derived based upon good statistical theory, which means that we can use them to develop standard tools that follow well defined rules of mathematics and logic. This makes them insensitive to opinion, feelings or subjectivity. We thus have the first crosslink in our bridge between the underlying population and a probability space with which we can associate our results.

A problem arises when you try to measure an infinite number of values in an infinite number of subjects and assign them to an infinite number of intervals. It is impossible to measure the density of all the igneous rocks, the blood pressure of all the hypertensive rats, or review all the hemoglobin assay results from a target laboratory, collate them into an infinite number of intervals, and still have time to report your results. You must draw a finite sample from the underlying population and generalize your results from the smaller cross-section back to the whole. The connection between *the sample* and the underlying population forms the second crosslink in our bridge.

The theory we are about to explore, and the tools we use to exploit it, require the linkage between the underlying statistical population and the sample to be undistorted. We gave you one example earlier about how a statistical population, tail lengths, yields misleading results when misapplied to a target population, hypertensive rats. That was a case of blatant silliness. But an even more insidious kind of error could creep into the process which could yield similarly misleading results yet remain almost undetectable. Suppose you are interested in a target population composed of all heart attack survivors, and suppose you sample patients from your local veterans hospital. The first problem is that you will probably skew your results to mostly men. In the USA, the majority of veterans hospital patients tend to be men in a lower than average socio-economic group, and your chance of observing a truly representative sample of heart attack victims is therefore minimized. Depending upon your geographical limits, you may be excluding population members from other parts of the country who would contribute valuable information to your study. If you are working in a rural area, all your patients may be from small towns or farms, or people who otherwise lead an entirely different lifestyle to that of a New York City stockbroker, or a Chicago taxi driver. Choice of sample is very important: you could easily bias your results by choosing your subjects too selectively, what we call selection bias.

Intuitively you already know what selection bias is: something in the selection process somehow favors the choice of one particular subgroup over another. To the statistician, the term *bias* has a very specific meaning: formally, any factor which interferes with the connection made between the target population and the sample is called *a selective factor.* The effect of all these factors taken together distorts this connection and enhances the differences between these two very important populations: the conglomerate effect is called *bias.*

A word of caution: to the classicist, the term *sample population* is a misnomer and oxymoron. A sample cannot be a population since it is not infinite or complete. But to help you understand the text more clearly, we will use this term intermittently. We think that by saying sample population, you will more readily see the connection between things you want to describe, such as all the hypertensive rats in existence, and the ones you can get your hands on, the six individual rats in your laboratory.

The theory developed to associate sample and population depends upon a minimum of distortion, which can only be ensured if your subjects are selected randomly from the underlying population. The act of randomization ensures that every subject has an equal opportunity of being selected for the sample without bias or interference. This is actually an exercise in mechanics: each subject must be given an absolutely equal chance of participating in your study. Assigning subjects to a treatment group in a laboratory is a lot easier than sampling the human population in a clinical trial, but the theory remains the same: randomization schemes using random number tables (or random number generators, etc.) ensure fair and honest sampling. Randomization of experiments and the identification and control of bias are discussed in more detail later.

Finally, suppose you were to carry out your experiment many times. Do you really think you would obtain the same results from sample to sample? If identical results were obtained, surely, as a good scientist, you would be at least a bit skeptical about their validity? We all know that variation between experiments exists, and we expect to see it. If we do not, we feel a bit uneasy about the validity of our study. Such variation arises from the fact that when you draw a finite number of subjects at random from your infinite underlying population, the chances of selecting the same subjects in different samples are infinitesimally small. We **should** see variations from sample to sample. The point of statistical analyses, in general, is to quantitate the degree of variation we can reasonably expect, and the point of descriptive statistics, in particular, is to provide an insight into the shape and size of the signal underlying your sampling noise.

MEASURES OF CENTRAL TENDENCY

Terms you should learn:

Mean (true)
Median
Mode
Sample mean
Random variable

Concepts you should master:

Limits of the median and the mode
Random variables, functions, and distributions
The sample mean as a random variable
Central tendency as a measure of location
Sample mean as an unbiased estimator of the true mean

Suppose you are allowed 5 minutes in which to discuss the results of your last six studies. Or suppose you must write a short communication summarizing these results for a prestigious journal. How do you communicate, quickly and effectively, the key points of your work so that you will win your Nobel prize, obtain your promotion, etc.? What key elements of your study do you want to describe in the clearest fashion? Do you really want to outline every single subject in your target population, one by one, or could you present some summary to make your points clearly and efficiently based on your sample?

Although on rare occasions you really might want to describe your study on a subject-by-subject basis, most instances require discussion of a conglomerate effect, results being summarized using one or two simple descriptors derived from a sample of your statistical population. These measures need to be clear and concise, and they are hopefully representative of what the underlying statistical population is actually telling you. Although many measures are available, and we will discuss some of them below, the one used most often to summarize a sample data set is the average.

The average or mean

Statistically, we refer to the average as the *arithmetic mean,* or just the *mean,* or the *expected value,* and there are many good mathematical reasons why it should be used to summarize your statistical population. It is stable, it is usually unbiased, and it takes advantage of a rich underlying mathematical theory which allows us to make statements about the underlying population even though we have only sampled a small segment of it. We humans like to know what the typical patient, rock or rat looked like, felt like or weighed. For us to make decisions, whether they are related to medical interventions or to consumer products, it is usually sufficient for us to know how a population, on average, would be affected by our intervention. How much, on average, does the typical man weigh? What is the average density of steel bars coming off an assembly line? What is the average blood pressure of 70-year-old men?

We assume that characteristic measures of a population are reflected in the average population member, and that the average calculated from our sample actually represents the average value that would have been observed if the entire underlying population had been observed. In statistical terms, what we are saying is that the sample mean is an unbiased estimator of the true mean. In experimentation, industrial design, and even in recreational activities we adapt to these measures. We perform clinical trials to see whether the average patient improves after therapy. We build automobiles to fit the average body, and we can use averages as a measure of performance in sports.

The mode

The average is only one summary variable that describes the typical behavior of a population, i.e. the 'center' of a sample, and helps us locate it in your measurement space. The primary variables which summarize the 'center' of your sample are the mean, the median and the mode. As a group, these are called *measures of central tendency*. The easiest of the three to understand, the one that lends itself to pure intuition, is *the mode*. Recall the frequency distributions outlined above: the mode is the most frequent value attained in your sample population. No calculations or formulae are required to find it: you simply count your data and plot it. The problem with the mode is that while it tells you about your most frequently observed values, it tells you nothing about the rest of your sample, and hence the statistical population underlying it. A great deal of information is therefore being discarded. This problem is illustrated in Figure 1. The frequency distributions shown in parts **a** and **b** of the figure have the same mode, yet these two distributions clearly represent different underlying populations. This single descriptor is insufficient.

A second problem arises when a frequency distribution has two or more peaks – what a statistician calls a bimodal or multimodal distribution. What does the secondary peak mean? Could it represent another underlying population, or is it just a fluke of sampling and nature? A classic example is the mean arterial blood pressure measured in 'healthy males'. The frequency distribution sometimes shows a secondary peak at the higher end of the scale. One explanation has been that a subsection of the target population has essential hypertension, and this group emerges in some samples when blood pressure is used as one of the attributes defining 'healthy'. In fact, there are actually two populations involved in the sampling: a normotensive population and a population of individuals who have coped with essential hypertension. In this case the label 'healthy' actually means 'asymptomatic'. There is nothing magical or mystical about this example. Bimodal distributions can be observed all the time. The point is that the secondary peak may indicate that your measure and your selection factor, e.g. 'low mean arterial blood pressure' equals 'healthy', are confounded and overlap.

One final problem with the mode is that it implicitly depends upon the scaling, precision and accuracy of your measurements. Figure 2 illustrates this by considering a population that is measured four different ways. First, suppose 100 people are standing in a field. You fly over them in an aeroplane and measure their heights with your altimeter. The precision of your measure classifies your subjects into 10-foot intervals. Clearly you have a mode in the group from 0 to 10, with no data in 10 to 20, 20 to 30, etc. This is shown in panel **a** of the Figure. What does this mean? All you can say is that there are no giants in your population.

You then use a measuring stick which is exactly one foot long to measure each person in the field to the nearest foot. The results are shown in panel **b**. Your distribution has no one in the intervals 0 to 1, 1 to 2, 2 to 3, or 3 to 4. Some small number of people are assigned to the interval 4 to 5, most to the interval 5 to 6, some to the interval 6 to 7, and none to the interval 7 and above. Your mode is in the interval from 5 to 6 feet. You now know that your population contains no dwarfs.

When you discover that your measuring stick actually has 1 inch gradations on the other side, you re-measure your sample population to the nearest inch. The results of that measure are shown in panel **c**. Clearly the mode is emerging in the interval 5 feet 7 inches to 5 feet 8 inches.

A world famous nuclear physicist then tells you that she can measure your sample to the nearest 0.000001 inch. The new distribution has no mode at all. Figure 2**d** shows the distribution over your whole range of values. The intervals are 0.000001 inches long, and no interval has more

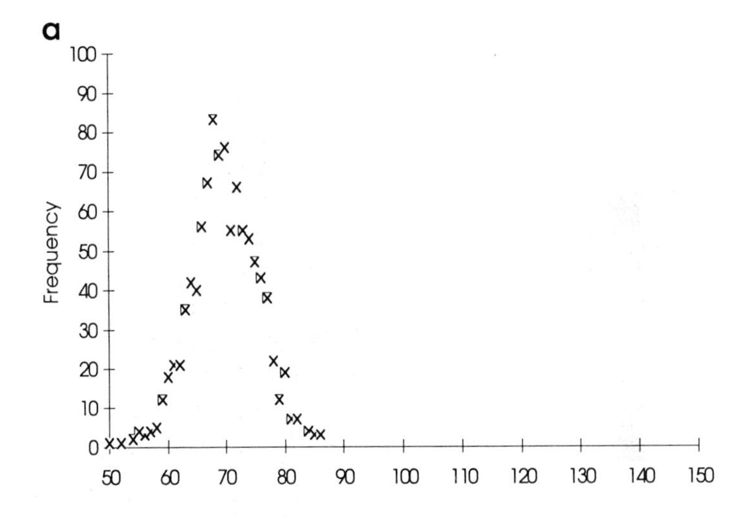
a
Frequency
100
90
80
70
60
50
40
30
20
10
0
50
60
70
80
90
100
110
120
130
140
150

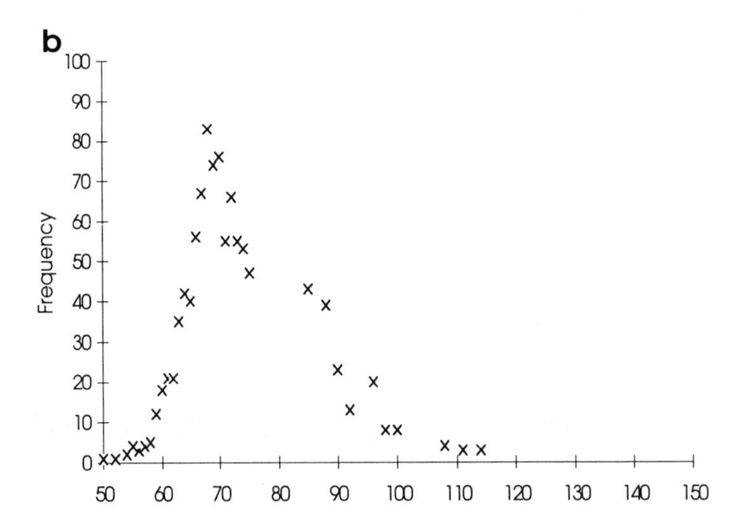
b
Frequency
100
90
80
70
60
50
40
30
20
10
0
50
60
70
80
90
100
110
120
130
140
150

Figure 1

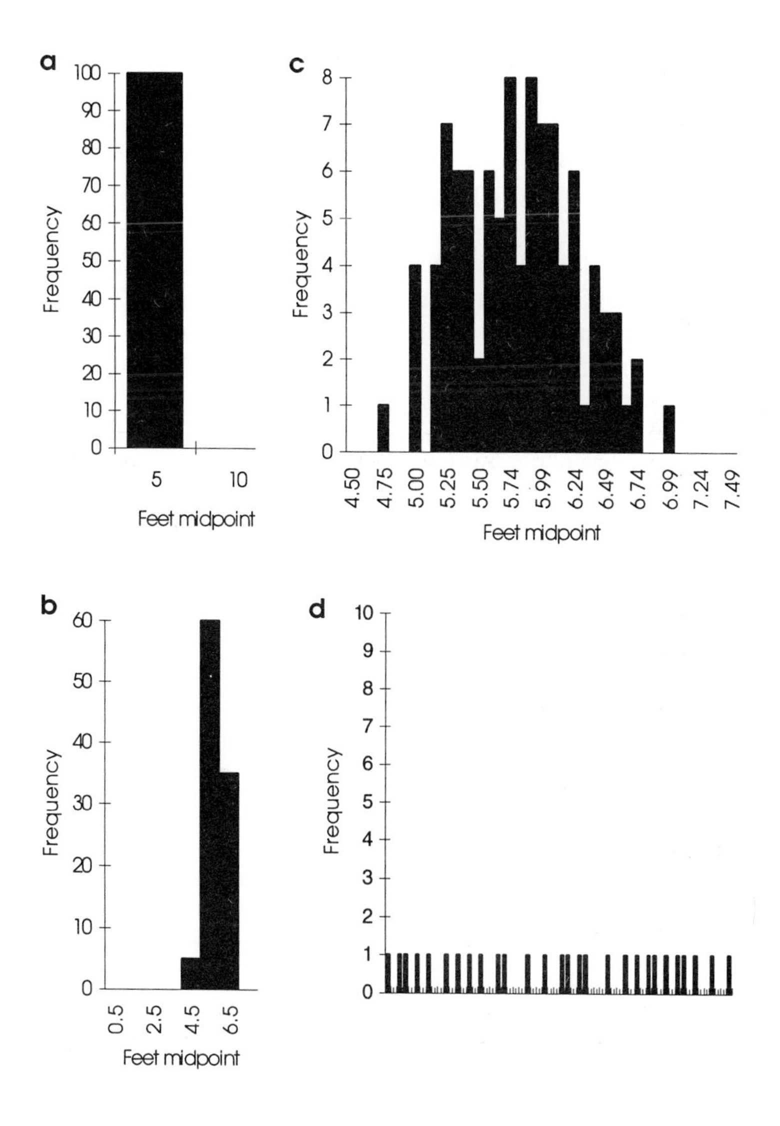
a
Frequency
100
90
80
70
60
50
40
30
20
10
0
5
10
Feet midpoint
c
Frequency
8
7
6
5
4
3
2
1
0
4.50
4.75
5.00
5.25
5.50
5.74
5.99
6.24
6.49
6.74
6.99
7.24
7.49
Feet midpoint
b
Frequency
60
50
40
30
20
10
0
0.5
2.5
4.5
6.5
Feet midpoint
d
Frequency
10
9
8
7
6
5
4
3
2
1
0

Figure 2

than one person in it. In other words, no two people in your sample are exactly the same height when it comes to that level of precision in your measurement, and although the data are infinitely precise, we can derive no useful information about the mode.

The median

Another measure of central tendency is the *median*. This represents the value that exactly splits your population in half: 50% of the population falls above this value and 50% falls below. Like the mode, the median is determined from the frequency distribution, and no real calculations are involved in determining it. Suppose you sample your underlying statistical population by drawing N subjects at random. If N is an even number, then the median is half way between the two values ranked $(N/2)$ and $(N/2) + 1$. If N is odd, the median is the $(N+1)/2$ ranked value in the data set. A characteristic of the median is illustrated in the following examples. The median of a data set comprising the values 1, 1, 3, 5, 6, 8, 10 and 100 is 5.5, half way between 5 and 6. If the values of the sample data were 1, 1, 3, 5, 7, 8, 10 and 100, the median is 6, half way between 5 and 7. Now suppose your values were 1, 1, 3, 5, 6, 8, 10 and 100 000. The median goes back to 5.5 again, illustrating the limits of the median when describing data.

The median intimately depends upon the sparseness of your sample but not the precision of your measure. This is one advantage of the median over the mode. There is another strength to the median in that it is insensitive to vastly outlying measures: the first and third examples above have the same median, even though the last point in the third sample is 1000 times larger than that in the first. This ability of the median to withstand outlying data measures is the basis for much of the non-parametric statistics described later and is one way in which statisticians handle potential outliers like those observed above. The problem is that there is no mathematically accessible way to associate the median of the sample with the median of the underlying population. To make that connection we use the third, and most common, measure of central tendency, the mean (the arithmetic mean or common average) described earlier.

Sample means and random variables

Everyone knows that the mean is calculated by adding the observations all together and dividing by the total number of observations you have. Let us look at the formula more closely:

$$\text{Mean} = \frac{1}{N} \Sigma X_i$$

$$= \Sigma \frac{X_i}{N}$$

Here Σ means 'summation'. What this simple formula says is that the observed value for the *i*th subject, X_i, is actually assigned an importance or weight in determining the final information derived from it about the statistical population: this weight is $1/N$. Each data point is, therefore, equally important in assessing the central tendency of the population, assuming that these points were chosen at random when the population was sampled, and that no data point is any more important than any other. It is this assumption that provides us with the theory we need to bridge the gap between our sample and the underlying populations.

Let us look at what the sample mean really is. It is derived from a finite cross-section of representatives selected at random from an underlying population. No data point is more important than any other. Even under these theoretically perfect conditions, would you really expect the average calculated from your sample to be exactly that of the statistical population underlying your experiment? Will they really be the same time after time and experiment to experiment? Is there no chance or randomness in your universe? Of course there is!

To link our sample to our underlying population, we use a mathematical construct known as the *random variable.* A random variable is actually a function. It represents all the possible values you could have chosen from your underlying statistical population if you had had the time. Each sample point observed represents one realization from all those possible choices. To understand how a random variable helps us link the sample to the underlying statistical population we need to lay a little theoretical ground work.

Every random variable follows two simple rules. The first is that anything you do to a random variable, whether addition, subtraction, multiplication or division, results in another random variable. The second is that each random variable is related to an underlying statistical population with a distribution of its own. Therefore, if a sample of size 10 is drawn from an underlying population, the resulting data points actually represent 10 realizations of the random variable. If we add them all together, the sum is still a random variable (rule 1). If we multiply that sum by 0.1, i.e., divide it by 10, then the result is still a random variable (still rule 1). This means that the sample mean (adding up your observations and dividing by 10) is a random variable, and the one sample mean you happen to have calculated is only one realization drawn from the distribution of all possible samples means calculated from samples of size 10.

We cannot make this leap with the median and the mode. While they are important descriptors which provide us with insight and an intuitive notion about the shape of our data, the mean is the primary tool for establishing an association between a sample and the population from which it is drawn. We will complete this connection in the next section when we talk about data dispersion and spread, but for the moment you can gather useful information using the three measures of central tendency already discussed.

Distributions

If we have a symmetric, unimodal distribution, then in the underlying population the median, mean and mode have the same value. The normal or Gaussian distribution is the primary example of this situation. As we diverge from symmetry the median and mean separate, since the mean is sensitive to outlying or extreme values and the median is not. As we diverge from a unimodal shape multimodal distributions arise, and while these may be symmetric, and while the median and mean may be perfectly equal, the mode can become totally undefined.

Given these three measures of central tendency, you should now be able to describe the vast majority of your experimental results in standard nomenclature to both your audience and yourself. Central tendency is, however, insufficient as a descriptor of your population. It only tells you about the location of your population, not its variability. To obtain a clearer picture of your data, to hear what it is really telling you, you need to know how variable it actually is. The next section provides us with the tools we need to describe that variability, and, on the way, helps us to complete the bridge between our sample and statistical populations by providing us with a measure of how good an estimate our sample mean really is.

Normal and non-normal distributions

Data may be drawn from an underlying distribution which is skewed (e.g. it has a long tail sticking out to the right). Distribution of this shape indicates that the frequency of occurrences about your mean is not symmetric. The weighting scheme outlined above for the arithmetic mean, which assumes that all points should be given equal weight, may therefore not be a fair representation of the importance of each data point to the final calculation. To offset this inherent inequality, statisticians have devised a way of transforming the data which works in most cases. If you converted the raw data values to their logarithms (assuming all are greater than zero) and plotted them, the distribution would appear much more normal. The antilog of the arithmetic mean of these log values (called the *geometric mean*) then gives you a better estimate of the true 'center' of the population. We discuss data transforms and scaling later (see The Design of Statistical Experiments). But the point here is that when faced with non-normal data, especially skewed data, there are tools on which we can rely that allow us to describe accurately and precisely those data and the underlying distributions from which they were drawn.

DATA DISPERSION, NOISE AND ERROR

Terms you should learn:

Variance
Standard deviation
Standard error
Kurtosis
Skewness
Interquartile range

Concepts you should master:

Measures of inter-subject distances
Measures of subject to point distances
Sum of squared distances and the insurance of a positive measure
Degrees of freedom
Units, square roots and Pythagoras
Data and scales
Distribution of sample means

In the previous section we outlined the three primary measures of central tendency you can use to describe your data. These measures tell you approximately where your data are, but not how the individual measures are arranged around them. For example, the two populations shown in Figure 3 have the same means and medians, yet they are clearly nothing alike! To truly describe your data you need a description of how variable your data are: how widespread is the distribution? Is it symmetrically spread over your observation range? How flat or peaked is it? The point of this section is to introduce you to the tools that measure data dispersion and how to use them to help you interpret your own results and then present them as clearly and concisely to others as possible.

Suppose you have at your disposal all the measures of a statistical population (for instance, heart rates) that were drawn from an underlying target population (which could be rats). How would you describe the variability of your population if you were not constrained by logistics, logic or mathematical reason? Intuitively, you would want a measure of dispersion that either describes how far each point is from every other point, or how far each is from some fixed reference point. The idea of providing an array of inter-subject distances is too vast a task to even initiate. Even if the universe you are dealing with is small enough, e.g. heart rates of rats that weigh 245 g and have green eyes, and even if the data array were manageable, there is no clear way for the

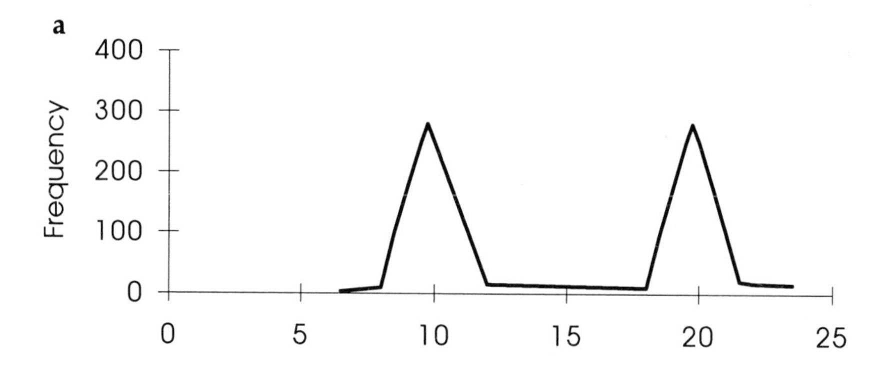
a
Frequency
400
300
200
100
0
0
5
10
15
20
25

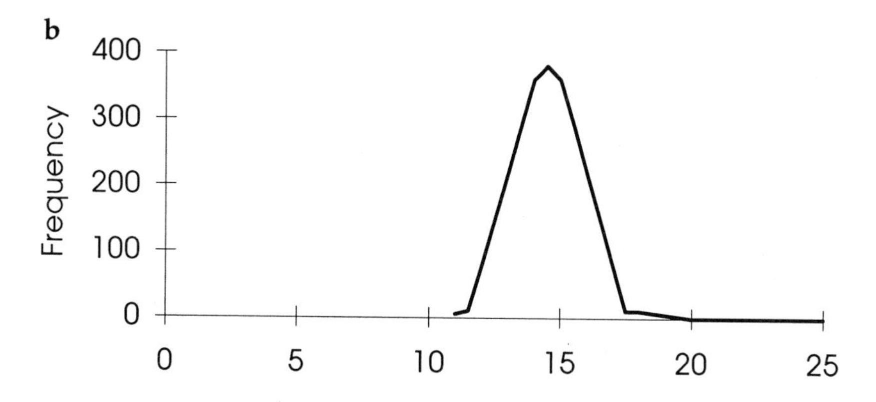
b
Frequency
400
300
200
100
0
0
5
10
15
20
25

Figure 3

human mind to digest this type of two-dimensional display and make good sense of it. There is certainly no good way to present these results to an audience in a concise manner.

You are therefore forced to measure the dispersion of your data around a fixed point – but which one? You could choose an arbitrary point in the data set, but then the measure of dispersion you pick would vary from population to population, making comparisons between them a nightmare. Alternatively, you could choose a fixed point that is independent for all data sets, such as zero, and proceed from there. The problem with that is that the measure of dispersion is now sensitive to the scale of the numbers. Data dispersed around numbers like 10 (e.g. 9, 11, 12, 14, 8, 7) will look less variable than those dispersed about numbers like 100 (e.g. 99, 101, 102, 104, 98, 97). It therefore makes more sense to choose one of the measures of central tendency described above to act as a locator, forcing your measure of variability to be centered in the correct location. Which measure would you choose? We have already shown that the median and mode are reasonable descriptors of location, but have problems associated with connecting the sample and underlying population. Since you will eventually be taking only a sample from your underlying population, you need to know that that linkage will eventually be in place. Therefore, the best choice for fixing the 'dispersion center' of our new measure is probably the mean.

Variance

Having chosen the mean as a fixed point, how would you establish a measure of dispersion that is a reasonable descriptor of the variability in your data? You could list the distances from each point to the mean in a table, but this is just a one-dimensional analog of the two-dimensional array of inter-subject distances described above. While it may seem easier to collate and display than a two-dimensional table, the information you can offer based on this table diminishes rapidly as your population expands. You should therefore provide your audience with some sort of summary measure: a single number that can be normalized across all underlying populations and can be used as a standard descriptor of data variability from population to population.

The obvious choice would be the average distance from each point to the mean – but there is a problem with that very simple solution. Look at the following equation:

$$\Sigma\left(\overline{X} - X_i\right) = \Sigma\left(\Sigma \frac{X_i}{N}\right) - \Sigma X_i$$

$$= N\left(\Sigma \frac{X_i}{N}\right) - \Sigma X_i$$

$$= \Sigma X_i - \Sigma X_i$$

$$= 0!$$

The left hand summation in the first line is the sum of the distances from each point to the mean. Dividing that sum by the number of points in the data set, N, would give you the average distance you desire. The solution to the equation, however, is zero. This makes good sense when you think about it. What this calculation tells you is that no matter what fixed point you choose, the mean is the single point that minimizes the sum of all the distances (if you do not ignore the sign of the distance), and that this minimum value is, by definition, zero. This is another advantage of the mean over the median and the mode. The conclusion that the center of your data is where that sum is not just minimized but must be zero is useful in developing the theory we will need to link the estimates from our sample population to the true underlying values of our statistical population. Since the sum is zero, and you know that there is noise associated with the data, your intuition has to stretch a little farther to produce a dispersion measure that will still work.

Intuitively, the easiest thing to do is to force all the distances to be positive. In other words, it does not matter whether a point is five units to the left or seven units to the right of the mean: all the distances are added up and the sum, or its average, is presented as a single measure of variability. This measure is called the *mean deviance* of the distribution and it seems to fit all the criteria we have set out for it. The problem is that mathematicians have found that the function that makes all the distances positive, the absolute value, is mathematically difficult to handle for reasons that are well beyond the scope of this book.

There is another way of making all these distances positive: they can be squared. Adding the squared distances between each point and the mean always results in a positive number. If you were to take that sum (called the *sum of squares*), and divide by N you would have the average dispersion about the mean. However, whether you realize it or not, you have forced a constraint upon your data: by using the mean as your measure of central tendency you have predetermined that the sum of the unsquared distances is zero. Therefore, if you knew the value of any $N-1$ points in your sample population, you could add them up and determine the final point explicitly. For reasons beyond the scope of this book, theory requires us to account for this constraint by dividing the sum of squares by $N-1$ and not N. This of course makes the ratio slightly larger, and thus our estimate of the noise a bit more conservative. It is this ratio:

$$\frac{\Sigma\left(X_i-\overline{X}\right)^2}{N-1}$$

that statisticians call the *variance*: it provides us with a single measure of variability which is mathematically easy to calculate and, it turns out, theoretically pleasing when associating a sample population to an underlying population. Parenthetically, the practice of using the number of independent points minus the number of constraints imposed on your data is used throughout inferential statistics to determine the freedom you have to maneuver about your information. The statistician calls these free steps, *degrees of freedom*, a concept which will be considered in much greater detail in the next chapter. For the time being, think of degrees of freedom as a 'currency' of information, much like information coins to be spent when describing or comparing your populations.

Standard deviation

Let us look at the units of the variance: they are the squared units of the original variables. Feet become squared feet, kilograms become squared kilograms, etc. How can we possibly relate to a dispersion measure that is inconsistent with our original units? How could we plot it on the same graph? What does it mean to say we have so many squared inches of variability? The obvious solution to our problem is to take the square root of the variance and define a new measure of variability with the correct units. This new variable, the square root of the variance, is called the *standard deviation*. A standard notation has been developed for these two important parameters. The population variance is usually denoted as σ^2, and the population standard deviation is denoted as σ. The sample variance and standard deviation are denoted, respectively, as s^2 and s.

The standard deviation has a number of advantages over the variance. First, you can plot the data and its spread on the same graph. A standard deviation of 3.41 feet around a mean can be plotted as a line on a graph which has feet as the x-axis. It has much more meaning than the variance of 11.63 feet^2. Second, the standard deviation looks much like the Pythagorean theorem (the square root of the hypotenuse of a right-angled triangle is the square root of the sum of the other two sides), which is, if you think about it, a measure of distance on a two dimensional grid. The linkage of dispersion about the mean and distance is more than fortuitous. It allows us to associate distance and its probability measure as the basis of the inferential statistics we present below.

Coefficient of variation

There is still a potential problem in interpreting your results when scaling your data. Suppose your underlying population consisted of only hypertensive rats with green eyes, and there are only six of them in the entire world. Your statistical population is composed of body weights, and the body weights of those six rats are 280, 286, 310, 316, 293 and 306 g. The mean body weight of your population is 298.5 g. The variance is 1023.5 g^2 and the standard deviation is 31.99 g. Now suppose you measured the body weights in kilograms. Then the mean is 0.2985 kg, the variance is 0.0010235 kg^2 and the standard deviation is 0.03199 kg. Are these data, measured as kilograms, really less disperse than when they are measured as grams? Of course not!

To adjust for the scaling of data, you could normalize your measures of variability by dividing by the mean. Dividing the variance by the mean makes no sense, since you still have a disparity in your units of measurement, but dividing the standard deviation by the mean gives the *coefficient of variation* (abbreviated CV), which is usually reported as a percentage. In our body weight example of green eyed hypertensive rats, the coefficient of variation is 0.107 (31.99/298.5 or 0.03199/0.2985) or 10.7%: that measure is the same whether body weight is measured in grams, kilograms, or metric tons.

Standard error of the mean

Up to now we have focused our attention on describing the underlying population. Suppose, however, that you have only a sample (chosen at random) from your underlying population. This is the scenario most likely to be encountered in the real world. Can we use these measures of variability to finally build our bridge between the sample population and the underlying population and link their respective descriptors?

Recall that a mean calculated from a sample is, in fact, a random variable. By its very definition, then, it represents a small cross-section drawn from an infinite population of other sample means, i.e. your particular sample mean is but one realization drawn from a distribution of all possible sample means when the sample size is fixed to N. This distribution of sample means has its own variance, and we would expect the realizations drawn from this distribution, of which your particular sample mean is but one, to vary from sample to sample. This becomes intuitive when you consider the chances of drawing exactly the same subjects and measuring exactly the same blood pressures from experiment to experiment. To link the sample and underlying populations, to help us complete our bridge, and to be confident that the first really represents the second, we must be able to measure how noisy the data really are, and how much of that noise leaks into your samples from experiment to experiment when you calculate your sample mean.

The tool used to gauge the ability of a single sample mean to estimate the true mean is the standard deviation of the population of sample means. We call this statistic the *standard error of the mean* (SEM). When a statistician talks about error, he or she is not making a value judgement. The term usually refers to noise in the data, and this noise may be due to limits in technology, subject-to-subject variations, day-to-day-variations, etc. The SEM is calculated from the equation:

$$\frac{\sigma}{\sqrt{N}}$$

where σ is the standard deviation of the underlying statistical population and N is the sample size. It is highly unlikely we would ever know σ, since our data are derived from a sample, so in practice, to calculate the SEM we substitute the sample standard deviation, s, for σ to derive the equation:

$$\frac{s}{\sqrt{N}}$$

Let us return to our green eyed hypertensive rats. Suppose there is no reason to believe that eye color has anything to do with body weight, and that these six rats have been chosen at random from the underlying population of all hypertensive rats. The target population is hypertensive rats, the statistical population is rat body weight, and the sample of six rats yields a random sample of all those possible weights, i.e. six realizations of the random variable 'body weight'. Then the sample mean, variance, and standard deviation are 298.5 g, 1023.5 g^2 and 31.99 g respectively. The SEM is 13.06 g (31.99/2.45, 2.45 is the square root of 6).

There are two things to note about the SEM. The first is that it also has the same units as the mean. The point of the standard deviation was to ensure that as a measure of noise it would have the same units as the mean, and the SEM is just the standard deviation of the distribution of sample means. The second point is that the SEM is smaller than the standard deviation, and that it becomes still smaller as the sample increases with size. The reason becomes obvious when you think about it logically. The SEM is a measure of variability for the distribution of sample means. If data are drawn from the underlying population at random, what is the likelihood that, all things being equal, all the measures will come from the extrema of the distribution? It is more likely that the majority of values will be drawn from the most frequently observed parts of the distribution. Even if this were not the case, the likelihood that they will all tend to be higher (or lower) than average is smaller than the chance of some being high and some low. When that average value for each particular sample set is calculated, it is thus pulled toward the middle of the distribution. Repeating the procedure for all the possible samples of size N results in a distribution of sample means which is less disperse than the raw data making up the underlying population.

There is still one more thing to notice about the SEM and its relation to the sample size: as N tends towards infinity, i.e. as you sample more and more of the underlying population, the error measure about the sample mean moves nearer to zero. The reason is simple. Using the same argument presented above, the more points you take at random the smaller are the chances that they will all be drawn from the extremes or from one side of the underlying distribution. A mean determined from a sample of 60 is therefore probably pushed more toward the center of the distribution than one calculated from a sample of six.

What do these decreasing SEMs mean? Since the sample mean is an estimate of the true mean, the greater the sample size, and thus the smaller the SEM, the more certain it is that the estimate of the latter is given by the former. As the sample size nears infinity (i.e. as you measure the body weight of every single hypertensive rat in the world) the estimate of the true underlying mean body weight becomes infinitely precise. The point is that while the standard deviation represents the spread of your underlying data, the SEM represents the precision of your estimate of the true mean of that set. Even though your SEM goes to zero, your underlying standard deviation, σ, is constant – i.e. your data remain disperse. Thus when trying to describe the variability expected from a particular population, the appropriate statistical tool is the standard deviation (or sample standard deviation). To describe an estimate of the reliability of the sample mean to estimate the true mean (and any generalization one could make from it), the correct tool is the SEM. They are not interchangeable.

Shapes of distribution curves

Finally, we present two measures that will help you describe the shape of a distribution, and a method of describing data dispersion in populations which are not symmetric or normal. Shapes are hard to describe in words. Terms such as steep, peaked, flat, etc. are too subjective to be used unambiguously: they modify other descriptive terms which may be equally ambiguous. That is why we depend so heavily upon graphical displays to show our data. We will present some of the more common techniques for data display in the next section.

However, it is sometimes necessary to forego a graphical description and describe your results using standard, well-defined terminology. We have described the tools which can tell us where data lie (a statistic that measures central tendency, like the mean) and how disperse they are (a statistic that measures variability, like the standard deviation), but we have no explicit tools to tell us about its shape. What we need to finish off our picture are measures of symmetry and sharpness. These two statistics are called *skewness* and *kurtosis*.

Skewness

Skewness is a measure of asymmetry: it measures how much a distribution slumps to one side or the other. In more statistical terms it measures how much more of a population is in one tail of the distribution (at one set of extreme values) rather than the other. Distributions which lean to one side or the other are called either *positively skewed* if the longer tail is to the right or *negatively skewed* if it is to the left. Skewness may result from data that are truly disparate, or it may result if the population underlying the sample is not homogeneous. Suppose you measure weights from rats of different ages, and the population contains more young rats than older ones. The distribution of body weights may be skewed to the left toward the lower values measured in the younger rats.

Kurtosis

Kurtosis is a measure of 'peakiness' or 'sharpness' in the distribution. It gives an insight into whether your distribution has a sharp peak (positive kurtosis) or is just flat (negative kurtosis). The former may arise because the underlying population is too homogeneous, e.g. mean blood pressure in healthy 25-year-old white male atheletes, while the latter is seen if it is not homogeneous enough, e.g. measures taken from all comers, independent of species, strain, age or sex. The real question you must ask is, "Is the sample I have drawn really representative of the underlying population I want to make a generalization about?" Both skewness and kurtosis can help you decide if that is the case.

Interquartile range

Once you have determined that your data are asymmetric or multimodal, the typical variance measure could be misleading. However, there are still ways of determining how disperse your data really are. One of the most common is the *interquartile range.* This is calculated by listing the data and choosing the two values that mark the 25th percentile (the point at which 25% of the data are below a given value) and the 75th percentile (the point at which 25% of the data are above a given value). Taken along with the median, these two points measure central tendency and spread, and give you a reasonable view of your data.

GRAPHICS

Terms you should learn:

Presentation graphics
Bar graphs and line graphs
Graphical annotation
Error bars
Exploratory graphics
Dot plots, stem-and-leaf, histograms and box plots

Concepts you should master

When you should use a graph rather than a table
Which graphic style to use with which data
How to use graphical annotation to embellish your graph
How to depict graphical range with error bars
Using graphical scale to achieve a uniform information scale
How to explore your data with graphs

'A picture is worth a thousand words'. We know it is a cliché, but this really is true. Because the data we deal with and the ideas we want to communicate can be so complex, the picture that we draw has to make sense. When you use the graphical tools available to you in most statistical packages you must ask yourself:

Who are you presenting to?
What are you presenting?
Why are you presenting it?

You must consider your audience. Did you plot your data to obtain an overview of your own experimental results (you are your own audience), or is this a graph meant for final publication and presentation when you deliver your Nobel prize winning lecture? The difference is more than just stylistic. What does your picture show that cannot be obtained from tables, words, etc.? Are the data being presented truly representative of the effect you see, or is there so much summary between you and the data that you lose the message along the way? It is easy to become so involved in the graphic process as to forget what you are really trying to do: inform and educate rather than entertain. Why make a graph of the data in the first place? Remember, data points are not information; a graph should provide a theoretical context that facilitates their interpretation.

A computer can provide many ways to represent data, but do any of them make sense? A graph is a pictorial representation of **relevant** experimental information. When constructed properly, a picture really is worth 1000 words, but like other forms of statistical information, a poorly constructed graph can be ineffective, and it might even be misleading. Using a graph is most appropriate when the **quality** of the experimental results embodies the relevant scientific information. Tables and other numerical summaries are more appropriate when the **magnitude** of the responses is important.

In our day-to-day endeavors, we use graphs either to present conclusions or to explore data. The following sections discuss the many forms of presentation and exploratory graphics, and provide some guidance about the many graphical tools at our disposal. Most of these are commonly available in the popular computer packages. In addition, we will address some special considerations in style: the goal is to create a reliable picture that will present the scientifically relevant information produced by your experiment clearly and precisely, and which will be easily understood.

Presentation graphics

There is a rich array of presentation graphical types from which to choose. Many of these can be accessed through popular spreadsheet software packages, such as EXCEL. A mock-up of the online EXCEL graphics menu is given in Figure 4. Several of the graphical types shown in the EXCEL Chartwizard display can be used to present your results, but choosing the correct graphical tool for the task should be made on the basis of: (1) the type of information being summarized, and (2) the conclusion you wish to communicate with your graph. We will discuss some of the most useful tools here.

Bar graphs

Bar graphs are effective for presenting relative frequencies (percentages) or the magnitude of an effect when data are collected from *nominal categories*. By nominal categories we mean classifications which are not ordered in any way, such as sex, hair color, or formulation.

Graphs of data collected from *ordinal categories*, i.e., groups that can be ordered, such as time or dose, will be considered later.

Consider an experiment in which weight gain was measured in male and female rats receiving one of three diets. We would like to observe which diet had the greatest (or least) influence on weight gain. Showing the data as a table (see Table 1) is reasonable but, as shown in Figure 5, a graphical representation is better.

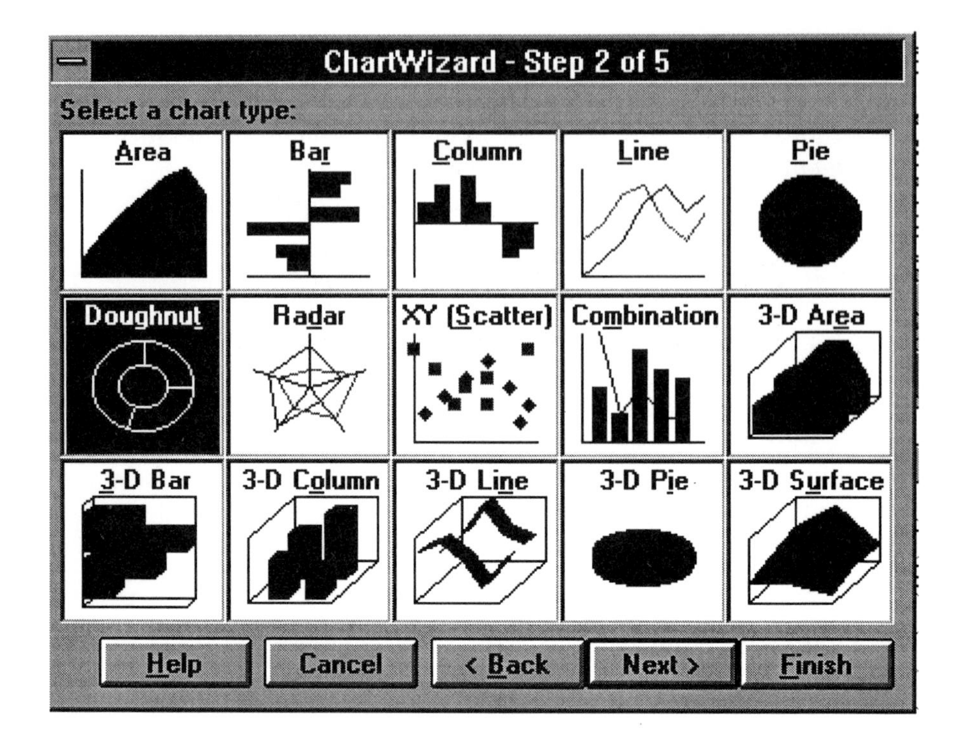

Figure 4 EXCEL graphics menu

Table 1 Effect of diet on weight gain in rats

Diet	Males	SE	Females	SE	Overall	SE
A	11	2.3	9	1.9	10	1.5
B	21	3.6	18	3.1	19	2.5
C	13	2	14	1.9	13	1.4

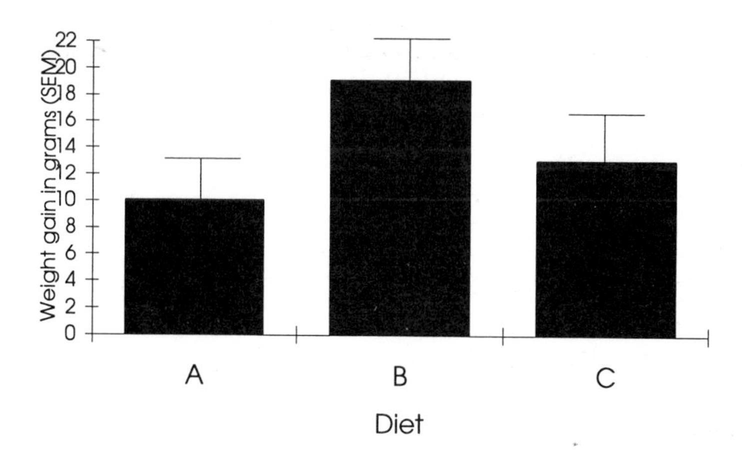

Figure 5 Bar chart showing weight gain in rats (data from Table 1)

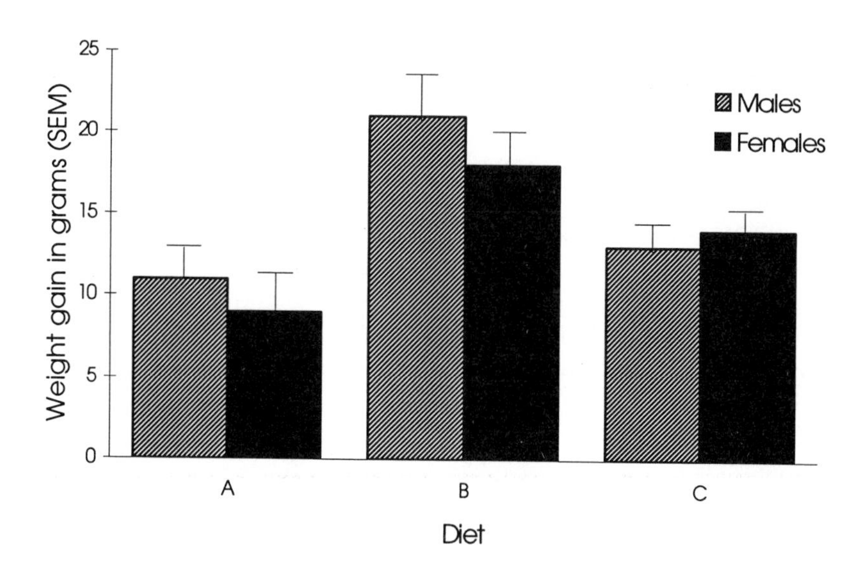

Figure 6 Weight gain in male and female rats (data from Table 1)

We choose a *bar graph* to depict the magnitude of weight gain for these three categories of treatment. The diets are nominal in that they are not ranked in any way: they merely divide the animals into groups. This plot combines the weight gain of male and female animals in the study together. If we wanted to see the responses of each of the sexes separately, we might elect to use a *grouped bar chart* (Figure 6). A grouped bar chart is useful for studying several *experimental strata* (in our example, sex) which are embedded in each nominal class (here, diet). This permits us to study simultaneously the relative differences in the effect of the study diets on male and female rats. Several adaptations of bar graphs are available in most popular graphical packages. These are less frequently used in scientific applications, but may be effective in special situations.

A *stacked bar chart* is a bar chart in which each bar is subdivided, proportionally, into areas representing your experimental strata. For example, suppose you were interested in total, rather than average, weight loss, and you wanted to see the contributions each sex made to that total. Total weight loss could then be plotted as a single bar such that the top of each bar represents the total loss for the males (hash marked) and the bottom half of each bar represents the total loss by the females (black). The height of the entire bar represents their combined total. The level of the division would be proportional to the total contribution made by the females to the total weight loss in each group.

This kind of chart conveys fundamentally different information to that shown by the grouped chart. It is used to evaluate simultaneously a total response and the contribution made by each subpopulation to that total.

A *3-dimensional bar graph* shows the magnitude of responses for two nominal classifiers at the same time. For example, suppose you expanded our simple example to include an exercise regimen in the study. Then, if there were three kinds of exercise and three types of diet your analysis of weight gain would be composed of 3×3 (=9) categories. In the third dimension would be the average weight gain per pair. You could then determine whether exercise regimen 1 varies more radically with diet than does exercise regimen 3. In other words, you can get a reasonable insight into any cross-classification effects you might observe with these factors.

Finally, a *pareto chart* is a special application of the bar graph, where the groups are ordered by magnitude of response. The pareto chart is commonly used to highlight significant effects in conjunction with factorial experiments. We will discuss these kinds of experiments in the section on Experimental Design.

Line graphs

Line graphs are used for presenting a trend over an ordinal measure, such as time or dose. The points that make up the x-axis are still classes but, unlike the nominal categories presented above, they carry with them an implicit order. For these kinds of data, line graphs are more effective than bar graphs for depicting the kinetics of response. Consider a stability study of two drug formulations which produced the data shown in Table 2. Here the form of the kinetics, perhaps the rate of degradation, is the relevant scientific information you want to communicate. The line graph in Figure 7 is the most effective way of depicting this attribute.

Table 2 Potency of drug formulations

	% of Initial Potency (SE)	
Time (Mos.)	Formulation 1	Formulation 2
0	100 (8)	100 (7)
3	70 (6)	80 (5)
6	40 (7)	70 (3)
9	30 (3)	50 (3)
12	10 (4)	70 (4)

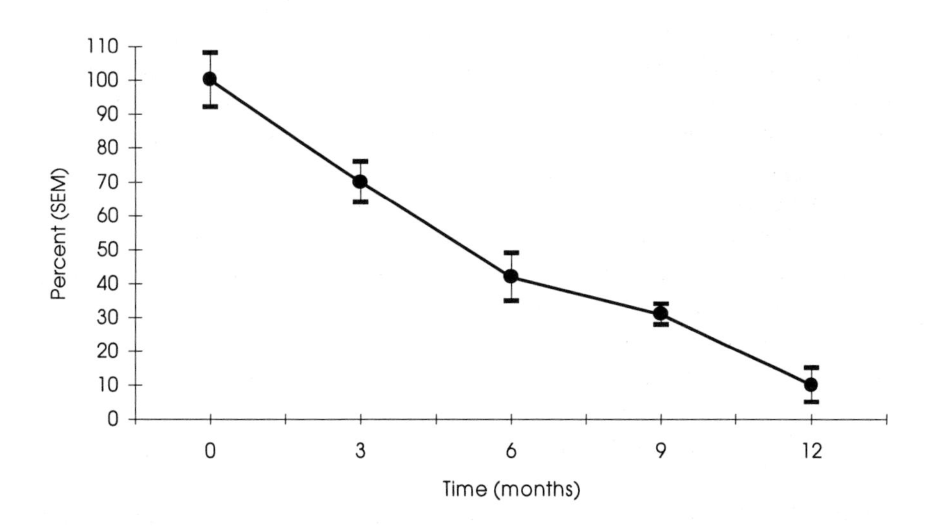

Figure 7 Kinetics of drug degradation (Formulation 1, Table 2)

Suppose you would like to compare the kinetics of several groups. The best way to do this is to add a second line to your graph, making it a *multi-line graph.* Contrast the effectiveness of a multiple line graph (Figure 8) and a grouped bar graph (Figure 9) for comparing the stability of two formulations.

The grouped bar graph in Figure 9 is too busy and detracts from the information. You are forced to look too hard to see the difference in degradation rates between the two formulations. You also cannot see the form of the kinetics. In this case, the line graph in Figure 8 is much more effective for depicting differences in the stability of the two formulations. This same distinction can be made for data collected over different doses of two drugs. Dose–response studies in pharmacology or safety experiments are best illustrated with line graphs or multiple line graphs rather than bar charts.

Several adaptations of line graphs are useful for portraying scientific information. A survival curve is a line graph. The data are derived from the proportion of subjects still alive at different times of the experiment. Sometimes these kinds of graphs are called *step graphs.* This particular tool is equally useful in depicting the proportion of subjects 'succeeding' where 'success' is associated with a known outcome. For example, the proportion of animals which improved after therapy, where you decide just what improvement is, can be plotted using a step line graph. *Control charts* are an effective tool for monitoring trends in scientific, analytical, and manufacturing processes. We will not discuss control processes in detail, but briefly, suppose you were monitoring a manufacturing process such as 'building a widget'. You could measure the width of each widget as it left the assembly line. Tracking the control responses (the widths) over all your runs would allow you to detect any noteworthy trends, outliers or deviations in the manufacturing process. It is much easier to perform this test graphically than it is to scan thousands of raw numbers by eye.

Special considerations

Several special considerations will affect the effectiveness with which experimental results are communicated by your graph, and they may well add to the reliability of your presentation.

Graphical annotation

Popular graphical software packages such as EXCEL provide a wealth of graphical enhancement tools which, when used properly, can enhance

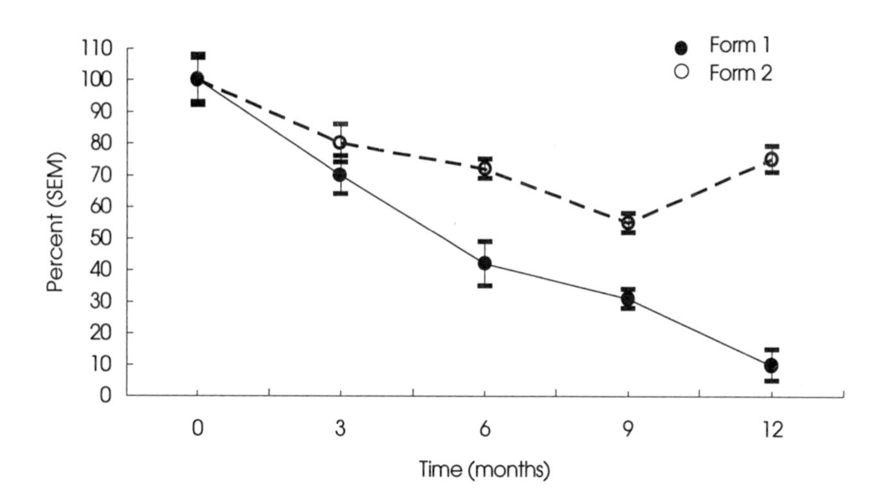

Figure 8

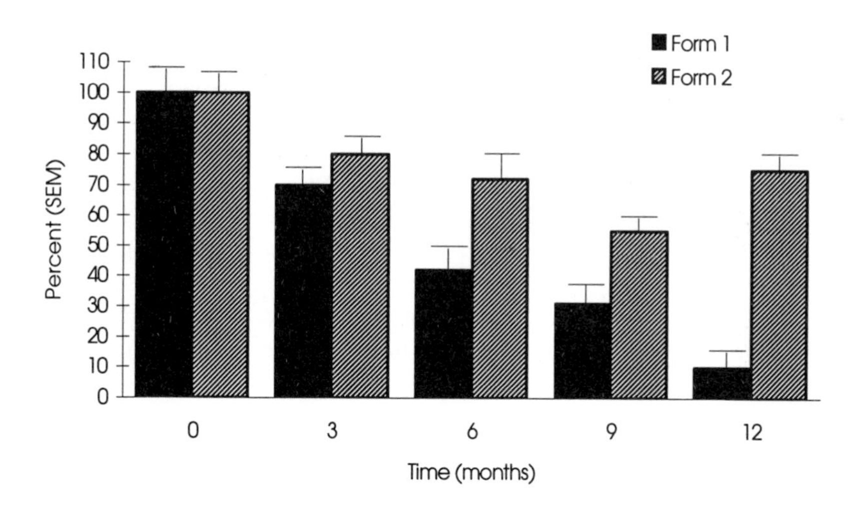

Figure 9

the information in your graph. These tools fall into two categories: labeling tools and referencing tools.

Labeling tools allow you to label the axes of a graph with legends, tick marks, and data labels. Graphical axes should be labeled judiciously. Labels on the horizontal axis should reflect the experimental conditions of your study, while those on the vertical axis depict the measurement characteristic. For the sake of completeness, the units of each (if appropriate, since some characteristics, such as hair color, have no units) should be included in the axis label. Any special characteristics of the graph, such as standard error bars, used in the graph can also be included in the axis description. Tick marks are used to illustrate design and scale: horizontal ticks can show the time points at which responses were measured or treatments were administered, while vertical ticks reflect the scale of the measured response. We will discuss graphical scale later, but without at least some indication of the scale on the graph your audience has no clue as to the size of the response. Finally, a legend should be included in your graph to delineate the experimental groups.

Referencing tools include symbols, colors, line styles, patterns and reference lines. These are used to add dimension to your graph. For example, different symbols can be used in a line graph to depict different experimental groups, e.g., circles representing males and triangles representing females. Different kinds of lines serve the same purpose. Different patterns can be used to fill the bars in a bar graph to highlight experimental strata as in Figure 6 above. Color is another useful way to add additional dimension in your graph. Thus, if different species of rat were included in the study of the three diet plans, two different colors and two patterns could be used to distinguish the two species and the sexes. Reference lines should be drawn at practically meaningful levels to draw attention to significant effects, but their overuse can be distracting and should be avoided. Some of these annotation tools are illustrated in Figure 10. In general, annotation is a powerful tool, but one which should be employed judiciously. It should be used to enhance the information value of your graph, not for entertainment: entertainment is a diversion, and the last thing you want to do is divert your audience from your message. If used incorrectly, these annotation devices will complicate your graph. Avoid redundant labeling and using too many group designators in a single graph. Popular graphical packages provide a host of clever annotation tools which, when misused, draw attention away from the scientifically meaningful information and detract from your presentation.

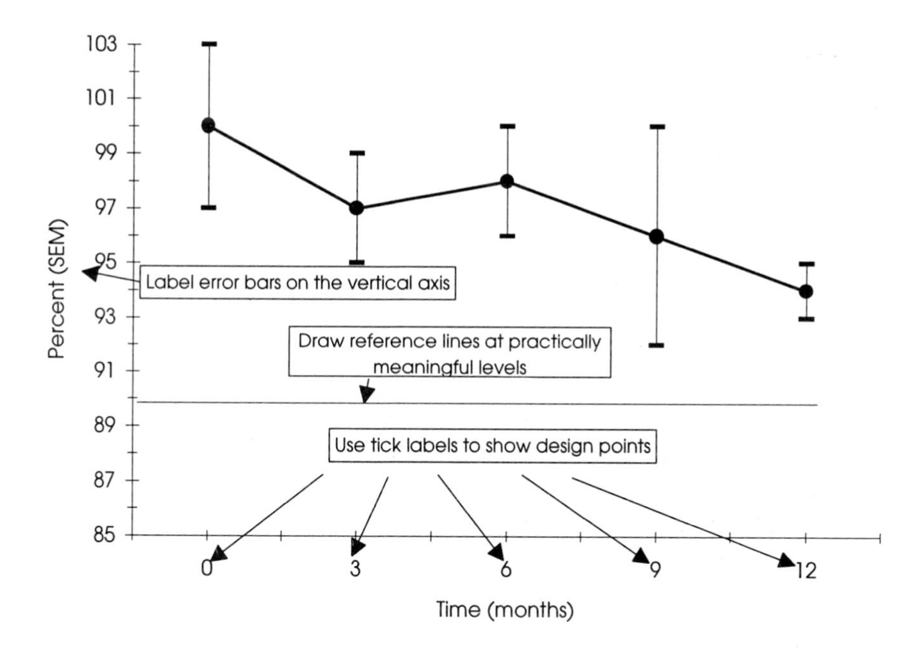

Figure 10 Annotation of graphs

Multigraphs

Multigraphs are a series of graphs which have been assembled on the same page. They are effective tools for both presentation and writing, and help to eliminate the need for flipping backwards and forwards through slides or pages. Multigraphs can be used to either reduce the burden of excess series on a single graph, or to plot several variables that might correlate well over time or dose. Thus, if you wish to track mean arterial blood pressure, heart rate, and glomerular filtration rate over time, and compare them side by side, they can each be plotted on individual graphs, then re-constructed into a vertical multigraph (i.e. stacked one on top of the other on the same page), displaying the opposing trends among the variables.

Graphical range and error bars

Graphical range can distort experimental information. An effect which is scientifically meaningful can be made to appear insignificant by selecting

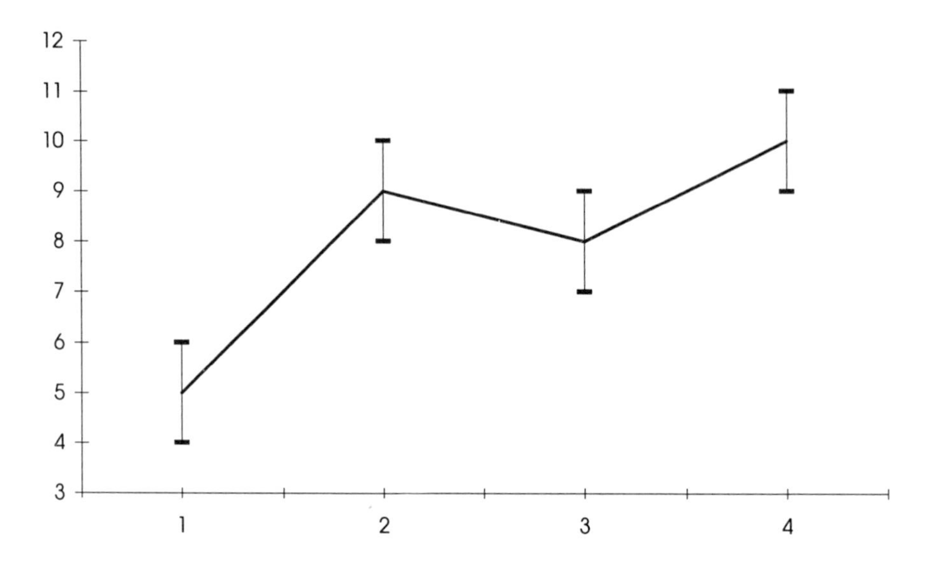

Figure 11 The use of error bars

a wide range, while a practically insignificant outcome can be made to seem large within a narrow range.

The range to employ in your plot should be selected, foremost, on the basis of scientifically reasonable limits. If mean arterial blood pressure typically ranges from 80 to 120 mmHg, then this should comprise the graphical range of your plot. Secondarily, the range can be fixed by the statistical quality of the information. Error bars are a measure of the statistical quality of the information, and can thus be utilized in lieu of a scientifically defined range (see Figure 11).

Graphical scale

Occasionally the scale of measurement is not uniform, and the quality of information varies with the level of response. Consider the pair of plots shown in Figure 12. Without the error bars, the difference between higher level groups appears greater than the difference between lower level responses. With the error bars the differences between groups 1 and 2 at the lower level and groups 3 and 4 at the higher level are quite comparable: data that have a uniform scale are easier to interpret. Changes can be explained quite impartially over the whole range of

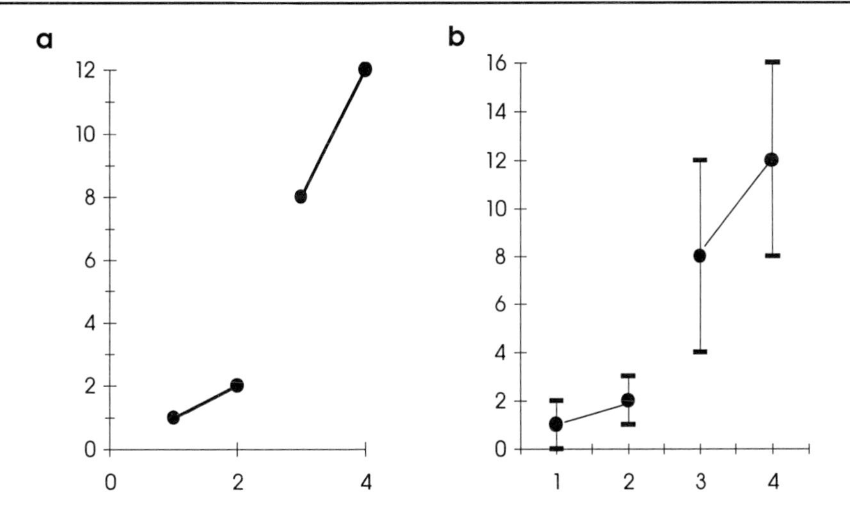

Figure 12 The same data plotted with and without error bars

responses. What would you do in this case? Alternative scaling may be needed to introduce a uniform information scale for measurements that vary in quality with the response. One common such alternative is geometric scaling (e.g. log scaling) which is frequently used to construct graphs of scientific results. Note, that summary measures used in a plot employing geometric scaling require special treatment. A *geometric mean and fold variability*, rather than simple average and standard deviation must be calculated for your display, otherwise, your graph will be inconsistent with the implementation of the alternative scale.

Dilution studies typically generate geometrically scaled measurements. An assay using a standard curve, in which test samples are diluted serially into the range of the curve is a case in point, as is a production process in which bulk material is diluted to final potency. Geometrically scaled data appear skewed when viewed in the linear scale. The larger numbers may even appear as if they are outliers, although in fact they form a natural part of a skewed distribution. Before generating a graph, therefore, be sure you understand the scale of your measurements. A class of graphical tools used for exploring the distribution of your measurements, and therefore their scale, is called exploratory graphics.

Exploratory graphics

Many of the statistical procedures discussed in this book are based upon the assumption that the underlying population of measurements follows a normal distribution. We have seen, however, that data generated in some scientific experiments can be skewed, and may even be contaminated with outliers. We would like to be able to explore our measurements for these characteristics before undertaking an analysis. *Exploratory graphics* let you study the distribution of a set of measurements graphically. Dot plots, stem and leaf displays, histograms and box plots are some of the tools available on many popular spreadsheet packages.

Dot plots

The distribution of measurements from experiments with small group sizes can be explored using an adaptation of a *scatter plot*, called a *dot plot*. A dot plot is constructed by plotting each treatment group on the abscissa (x-axis) and the response for each subject on the ordinate (y-axis). For example, consider an experiment in which urine output was studied in six dogs, each receiving either saline or a candidate drug. A dot plot of the results is shown in Figure 13.

In addition to serving as a graphical summary of the individual measurements obtained in the experiment, the dot plot highlights any unusual values which may be measurement outliers or artifacts of a non-symmetric distribution.

As noted above, the dot plot is a special case of the *scatter plot* or

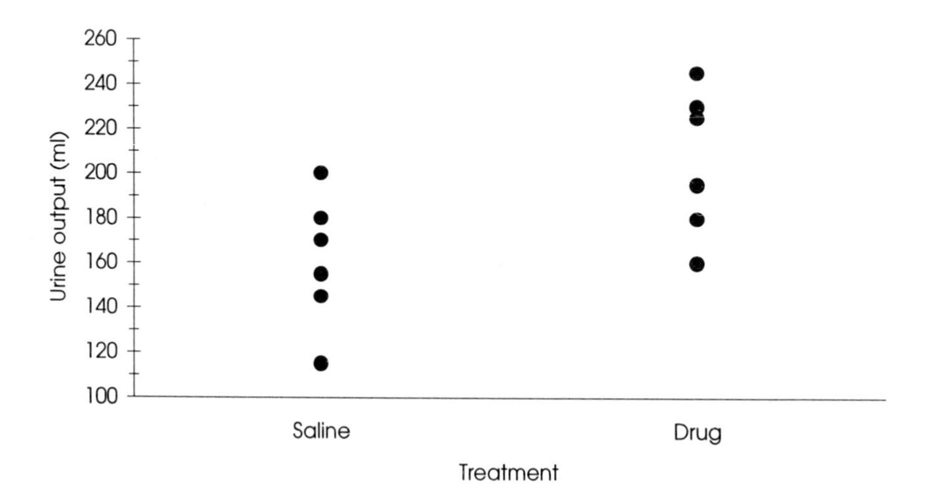

Figure 13 Effects of drug or saline on urine output in dogs: a dot plot

scattergram. In the case of the scattergram the x-axis is not divided into categories, but represents a continuous measure such as height or weight. The (x,y) pair from each subject is plotted as bivariate data on both continuous measurement scales. We typically use this kind of plot when we want to perform a linear regression or correlation. An actual scattergram of height vs. weight data is provided later in the Estimation section.

Stem and leaf displays and histograms

A dot plot is useful for exploring the distribution of measurements obtained from a statistical experiment when the number of observations is small (perhaps 20 or fewer). The utility of the dot plot diminishes as the number of observations increases, when the density of the plotted points obscures the shape of the data distribution. Exploratory graphical techniques which are available to study the distribution of a large data set are *stem and leaf* displays and *histograms.*

A stem and leaf display is a simple tally of a series of measurements and can be readily constructed with pencil and paper. Consider an experiment in which dissolution data have been collected on a particular formulation, yielding the following figures:

87	109	79	80	96	95	90	92	96	98
101	91	78	112	94	98	94	107	81	96

A stem and leaf display is constructed as follows. First, identify the 'stem' portion of your plot. In this case, the values are all factors of 10 (70s, 80s, 90s, 100s). The 'stems' are therefore 7, 8, 9, 10 and 11. List these values in a vertical column. For each observation, record the 'leaf' portion (the next digit in the measurements). Make a row of leaves corresponding to the appropriate stems (the higher order digits). The above example then becomes:

```
11 | 2
10 | 1 7 9
 9 | 0 1 2 4 4 5 6 6 6 8 8
 8 | 0 1 7
 7 | 8 9
```

The first row is the number 112. The next row are the three numbers 101, 107 and 109. The next row has 11 different entries.

The distribution of a larger set of measurements can be depicted using a histogram. This exploratory graphical tool is available in most of the popular spreadsheet packages, or can be constructed manually by dividing the range of measurements into six or more evenly spaced class intervals. Individual measurements are tallied into these class intervals, and a bar chart of the frequency (or the relative frequency, which is equal to the frequency divided by the number of measurements) is constructed. Figure 14 shows a frequency histogram of potencies for 487 batches of a product.

Although this plot appears to indicate that the distribution of potency measurements is skewed (shifted) to the left, i.e. some measurements appear excessively high, in this case it is a natural consequence of the inherent distribution and not due to outliers. It would be misleading to describe the normal range of these measurements using two or three standard deviation limits on the mean, since a higher proportion of readings would fall above the upper bound of the interval than would exceed the lower limit. In this case, a mathematical transformation of the potency measurements, perhaps a log transformation, would yield a

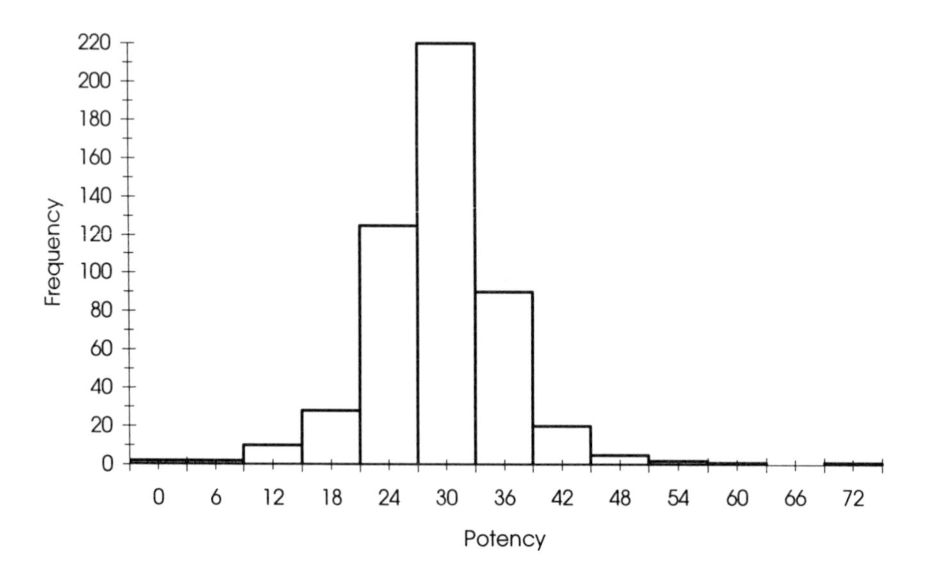

Figure 14 Frequency histogram for 487 batches of a product

more characteristic symmetric distribution. Statistical limits on the transformed mean will equitably capture equal proportions of high and low measurements.

Box plots

Another exploratory tool that is frequently available in scientific graphics and analysis software is a *box plot,* which uses percentiles of a set of measurements to depict the shape and range of the distribution (Figure 15).

The box is composed of a center line, which represents the median measurement. The upper and lower borders of the box represent the 'quartiles' of the distribution. Thus the middle 50% of the distribution of measurements falls in the range of the box. A skewed distribution might appear as a box with an off-centered median. The lines emanating from the box extend an equal distance from the median, and serve to identify outliers. The box plot outliers are noted as exceptional cases in the distribution and may either result from truly abnormal values or be the result of a skewed distribution. A box plot of appropriately transformed data (perhaps the log) might ameliorate this condition.

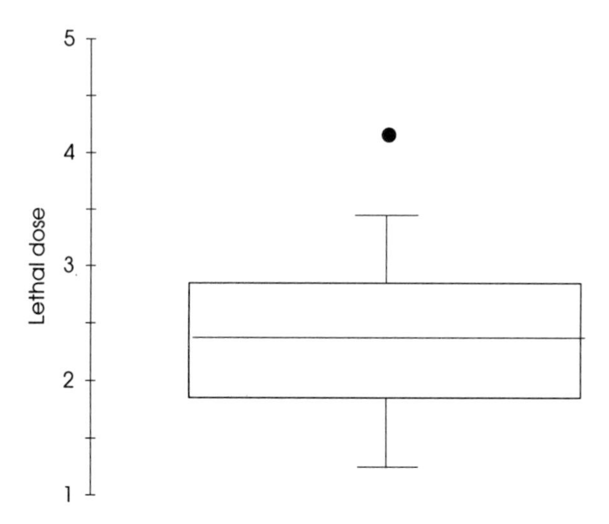

Figure 15 Box plots indicate the shape and range of a distribution

Summary

A well constructed graph is an effective tool for communicating the results of your scientific investigation, or for exploring your data. Like most statistical tools, however, a poorly constructed graph can distort the information and mislead your audience. The appropriate form of graphical display is dictated by the type of data collected and the message to be imparted. Bar charts are useful for displaying relative magnitude of responses across categories, while line graphs are more effective for displaying kinetics. Each is useful for studying relative effects in multiple groups. Exploratory techniques, such as dot plots, histograms, and box plots, are useful for studying your distribution. Most of the statistical techniques we use assume 'normality' in the distribution of measurements. Therefore, a symmetric distribution of sample measurements is the key to an accurate assessment of your experimental results.

2 INFERENCE

Terms you should learn:

Inference
Null hypothesis
Alternative hypothesis
p-value
Critical value
Type I error rate

Concepts you should master:

Distance in probability space
Conditional probability and the *p*-value
Critical value and risk
One-tailed and two-tailed test

When most people talk about biostatistics they are really focusing on the part of the toolbox we call *inference*. Inference is the process by which two or more groups of subjects are compared to determine, given the appropriate controls, whether an experimental intervention altered an outcome measure. We usually consider the average effect of treatment because, as we pointed out in our discussion of descriptive statistics, these measures provide simple summaries of complex population-based effects. If we can say that, on average, the blood pressure decreased, or the sodium concentration increased, etc. we are happy to conclude that the intervention was, for the most part, successful.

Every test (tool) in this part of our toolbox is based upon the following model. First, you assume that the intervention had no effect on the underlying populations. Samples from each population are compared to see how 'far away' from each other they are. The distance between samples (their means or other summary values) is then associated with a probability, calculated under the assumption that there really is no effect of your treatment. If the distance is 'large enough', that is, if there is very little likelihood that what you observed could have occurred by chance alone, then you can probably reject the underlying premise which assumed that the intervention had no impact on the system. All our statistical tests assume that our samples represent underlying populations, and that these populations are unaffected by the intervention being examined. This implies that, on average, the outcome values measured in a treatment group are the same as those measured in an appropriate control group. That does not mean that no individual showed an effect of treatment: although some subjects may have responded very positively (i.e. in the way expected) to the intervention, the group **as a whole** did not. Statistically, this assumption of 'no effect' is called the *null hypothesis*, and it represents the most conservative assumption that can be made about the underlying science of your process. The notation commonly used to represent the null hypothesis is H_0, and in each of the tests described below, we will explicitly state it in its mathematical parlance, and then describe, in plain English what it means.

Once you have assumed that the null hypothesis is true, you have to perform the next phase of the comparison. (Since most of our work is done with means, we will use the sample mean as an example of the summary measure we can test. Later in this chapter you will see that we can also test other measures.) This involves determining how far apart two sample means are when distance is measured in a probability space. The common measures of distance (feet, meters, light years) cannot be used to measure the distances between means: such distances are determined as a function of the noise or variance which surrounds them. Given that distance, we ask, 'What are the chances that the first sample mean is this far away from the second, assuming that there was no effect of treatment?' This probability is the *p-value.*

The *p*-value seems to be the Holy Grail of all experimental science. When asked whether a drug treatment worked, the typical response is 'Well, *p* is less than 0.05, so I guess it did'. The problem with that statement is that most people do not really know what the *p*-value is. The *p*-value represents the probability of observing a difference as large as that obtained (i.e. of being 'this far away' in noise units) **given the null hypothesis is true**. Statistically, we call this a *conditional probability*, and it represents the chance of rejecting the null hypothesis when, in fact, both samples were really drawn from the same underlying population: i.e. the treatment really does not work. The *p*-value thus represents your chance of making a mistake. It is not a measure of effect, but the risk you take of rejecting the null hypothesis **given the fact that it is true**. The point at which you are willing to take that risk is a function of how much you are willing to lose if you are wrong. If you are pursuing a screening protocol, and you know that a great many more assays will be performed before a final decision is made about the use of a given drug or intervention, you are willing to take a bigger risk than if you are performing a key assay that will determine the fate of a million dollar project. This 'comfort zone' about your risk of being wrong is called *the critical value* of your experiment. Typically it is chosen at 0.05, i.e. 1 chance in 20. Why? We do not know, but at some time in the past someone decided to accept an error rate of 1 time in 20 and risk the consequences. The statistician calls this risk of being wrong *the Type I error rate.*

If the null hypothesis is rejected we need *an alternative hypothesis* to fall back on. The statistician usually denotes this dichotomy as:

$$H_0: \mu_D - \mu_S = 0 \text{ vs. } H_A: \mu_D - \mu_S \neq 0$$

Statistically, we are saying that if we assume there is no real difference between the two treatment groups, i.e. the true mean of a drug-treated group, μ_D, is equal to the true mean of a saline-treated group, μ_S, then any difference we observe in the sample means is strictly due to the sampling distribution and chance alone. If the data observed are 'different enough' from zero then this underlying assumption might be rejected. Details on how to calculate this distance and associate it with a chance of occurrence are presented below.

If you reject the null hypothesis, you have a choice of two 'research hypotheses' which depend intimately on your alternative hypothesis. If an investigation is aimed at proving that drug treatment can **only** increase urine output in the treated dogs, i.e. our alternative hypothesis is that $\mu_D > \mu_S$, then *a one-tailed test* should be performed. However, if you are interested in showing that the two means are different, i.e. $\mu_D \neq \mu_S$, and you do not care in which direction the difference occurs, then *a two-tailed test* is performed. This decision should be made when you design your experiment, and should be based strictly on scientific first principles.

COMPARING A SAMPLE MEAN TO A POPULATION WITH KNOWN MEAN AND VARIANCE – THE ONE SAMPLE *z*-TEST

Terms you should learn:

Critical ratio
Standard normal curve
z-score or *z*-statistic

Concepts you should master:

What the null hypothesis really says
Critical ratio as a distance
The true meaning of the *p*-value in this test

Case study

Suppose you have access to the complete medical records of a population, and from these you know that the mean serum cholesterol level in your area is 237 mg/dl. Using the tools described above, you further determine that the distribution of serum cholesterols has a standard deviation of 20 mg/dl. You believe that a regimen of health foods, reasonable exercise and stress management will lower these cholesterol levels. The question you are asking is 'If I were able to control lifestyle with these regimens for three months, could I really decrease serum cholesterol levels below those of the normal population?'

To test this hypothesis, you choose 100 people at random from the general population and sequester these (willing) volunteers in a holiday resort for three months, ensuring that they are well fed, exercised, and kept cool, calm and collected. You then measure their serum cholesterol levels.

Table 3

Patient	Serum cholesterol after treatment (mg/dl)
1	206
2	194
3	221
4	196
.	.
.	.
.	.
99	200
100	187
Mean	209.22
SD	16.34

The data

The data collected from your experiment, along with preliminary summary statistics, are presented in Table 3 and shown as a histogram in Figure 16.

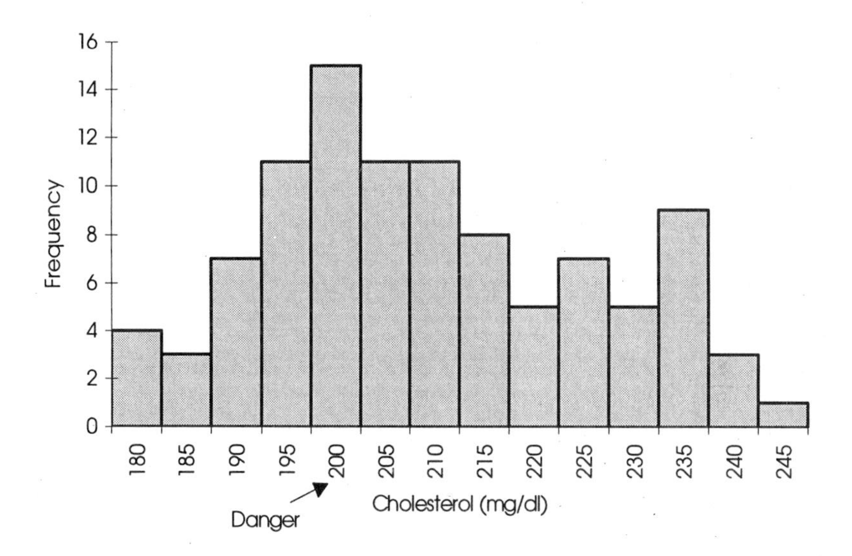

Figure 16 Distribution of cholesterol levels

Did you really lower cholesterol levels significantly below those of the general population?

Data analysis

You know that the true mean serum cholesterol for the general population is 237 mg/dl and it has a standard deviation of 20 mg/dl. (That is to say, the computerized records from your national health statistics center completely and adequately describe the general population in enough detail that these numbers are as close to 'known' as you can get.) The statistical question is 'What is the likelihood that the serum cholesterol level of 209.22 mg/dl measured after treatment was drawn from a population with a mean cholesterol level of 237 mg/dl and standard deviation of 20 mg/dl?' This question forms the basis for the null hypothesis.

Table 4 shows the answer calculated using standard analysis.

Table 4

	After treatment
Mean	209.22
SD	16.34
z-score	13.89
p-value	<0.001
Critical value (one-sided)	1.64
Critical value (two-sided)	1.96

We will discuss the calculation of these values later, but first notice that the table supplies all the information needed to enable you to decide whether your treatment is really as good as you think it is. After treatment, the mean serum cholesterol is 13.89 noise units away from the true population mean of 237 mg/dl (the *z*-score). The probability of observing means this far apart, **when the null hypothesis is true**, is less than 1 in 1000. Rejecting the null hypothesis when it is true is called a Type I error. Recall that we defined the Type I error rate as a comfort level for this kind of error, and that it is set before the experiment begins. The *critical value* is the number of noise units which fixes the Type I error rate to that comfort level. If you did not want to reject a null hypothesis which is true more than once in 20 times (i.e. 5%), the critical value which ensures this error rate is 1.64.

A quick note about the choice of the critical value. You assumed that your treatment would only decrease serum cholesterol levels. If these levels increased you would not care by how much, and, hopefully, you would discontinue the study immediately. When these conditions pertain, a one-tailed test is warranted. If, on the other hand, you were unsure of the effects your treatment would have on mean serum cholesterol, and if you wanted to detect any movement at all, then a two-sided test should be performed. In that case, the critical value would be 1.96.

In either case you can easily reject the null hypothesis for these data. You can thus conclude, with at most a 5% chance of being wrong, that you would not be able to observe a mean serum cholesterol of 209 mg/dl if your patients were drawn from a population with a true mean and standard deviation of 237 mg/dl and 20 mg/dl respectively. The *z*-statistic and the one-sided critical value are shown schematically in Figure 17.

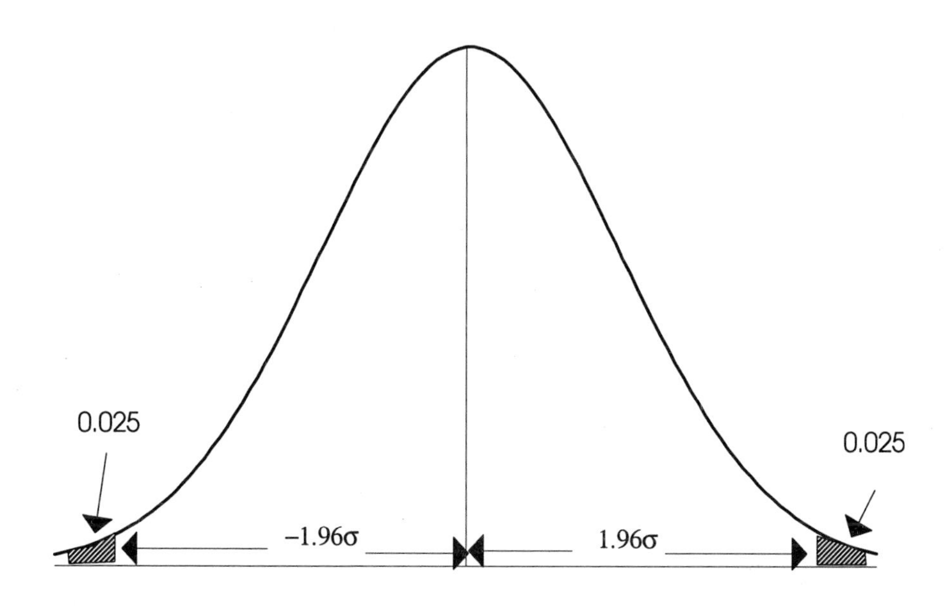

Figure 17 Gaussian distribution

The 'bell-shaped' curve depicts the *standard normal* or *Gaussian* distribution. The *z*-statistic, the distance between your sample mean and the known mean (as measured in noise units) falls well into the tail of this distribution. The likelihood of seeing a mean cholesterol level of 209 or less, given that the null hypothesis is true, is therefore exceedingly small (less than 0.1%), and falls well within the limits required to satisfy the 5% error criteria for declaring the null hypothesis false. The null hypothesis is therefore rejected and you claim that there is a 'significant' decrease in serum cholesterol levels after treatment comprising a long holiday, good food and exercise.

The question and your experiment

The study was based on a question about the efficacy of healthy food, healthy air and healthy exercise in maintaining cardiovascular health. The scientific target was the cardiovascular health of the general (target) population, and serum cholesterol level was taken as a reasonable predictor of cardiovascular well-being. Standard medical technology was used to measure serum cholesterol levels in the statistical population, and lower cholesterol level was assumed to be, in general, indicative of a healthier state. What the experiment is asking is, 'Is this treatment plan a healthy regimen for the general population to follow?', or to put it another way, 'On average, did this treatment really lower serum cholesterol levels below those of the general population?'

Assumptions

The scientific assumptions underlying the one sample *z*-test are:

(1) the true mean of the underlying population is known;
(2) the true standard deviation of the underlying population is known;
(3) the measurements taken reflect the true effect being looked for, i.e. data being measured are predictive of your proposed effects;
(4) the underlying population is truly homogeneous.

The statistical assumptions required for this test are:

(1) that your measures are continuous (e.g. not all-or-none effects, scores, etc.);
(2) that they are drawn from an underlying distribution which is normal;
(3) the sample population is chosen randomly from the homogeneous underlying population.

We will discuss these assumptions and what happens when they are violated below.

The test – the equation, what it means, and the distance it measures

The null hypothesis is given by:

$$H_0\colon \mu = \mu_0 \quad \text{or} \quad H_0\colon \mu - \mu_0 = 0$$

In English, this says that the mean of your underlying population is μ_0. In this analysis, we test to see whether you could observe a sample mean consistent with that assumption. In other words, 'What are the chances of observing differences between the sample mean and the true mean 'this big' given that the sample was drawn from the underlying population?' The statistical assumptions upon which this test is based actually restate part of this null hypothesis.

The equation which tests this assumption is given by:

$$z = \frac{\overline{X} - \mu_0}{\sigma / \sqrt{N}}$$

where $\overline{X}$ and N represent the sample mean and sample size, and μ_0 and σ represent the true mean and standard deviation of the underlying population.

This equation (known as *the critical ratio)* says that we are interested in a difference between the sample mean (a random variable) and a fixed point, the true mean. The equation also tells us that this difference (in mg/dl for the above example) should be normalized into a new distance (measured in noise units), and that the normalizing factor should account for the normal variations expected in the underlying population.

Think about the numerator of the critical ratio. The difference, Δ, is actually a random variable. The test asks whether a single realization drawn from the distribution of all Δ's is far enough away from the true mean Δ to cause us concern. Under the null hypothesis, the Δ distribution should have a mean of zero; although the difference between the true mean, μ_0, and a particular sample mean will never be exactly zero, how far away from zero is it, and could that difference have been observed by chance alone?

The only information left to be determined is the dispersal of sample Δ's around zero given the null hypothesis is true. Recall that we are asking what the difference in **mean** serum cholesterol levels is, and that the standard deviation of a distribution of means is actually the standard error of the mean (see Descriptive Statistics). The standard noise measure should, therefore, look something like a standard deviation divided by the square root of a sample size. When the null hypothesis is true, the standard deviation of the differences between any individual and the true mean ($\Delta_i = \mu_0 - X_i$) is given by σ. The standard error for the difference in means should, therefore, just be σ divided by the square root of N. If we normalize the difference observed by this standard error-like normalizing factor we derive the critical ratio. The critical ratio is also a random variable, however, and it must, therefore, be represented by a distribution. This new distribution is the tool we use to associate the noise units to the probability measure we are actually seeking. Remember the bridge we built in the discussion of Descriptive Statistics above? This new distribution also has a mean of zero but it now has a standard deviation of one because of our normalization procedure. It is called the *standard normal* or *z-distribution,* and is calculated from a complex mathematical formula which describes the bell shape shown in Figure 17. The probabilities we seek are the cumulative areas under the curve, i.e. the areas to the right and left of any arbitrary cut-off point. Any differences greater than a given cut-off, e.g. 1 noise unit, have a fixed probability of occurring when the null hypothesis is true. These probability measures have already been tabulated and appear in every standard statistical reference book; these are the tools we used to obtain our *p*-values in the earlier calculations.

Returning to our example

The calculations associated with our example data are outlined in Table 5. The test is performed as follows: first calculate the numerator for the critical ratio, i.e. the difference, which is 27.78 mg/dl. The denominator (based on σ and N) is calculated as 2 mg/dl. The distance between sample and known mean serum cholesterols is 13.9 noise units. To determine the probability that they are 13.9 or more noise units away from each other, i.e. to link the distance measure to a probability *when the null hypothesis is true,* we turn to the z-table and find that the likelihood of observing a difference this big is less than 0.001. The chance of wrongly rejecting the null hypothesis is much less than 5%, the error rate with which we appear to be comfortable. You can now conclude that serum cholesterol level is significantly decreased by healthy air, healthy food and exercise and, therefore, that this regimen is good for the cardiovascular system and, probably beneficial to the general health of the public.

Table 5

Formula	Example
$\Delta = \mu_0 - \overline{X}$	$D = 237 - 209.22$ $= 27.78$
σ is known	$\sigma = 20$
$z = \dfrac{\mu_0 - \overline{X}}{\sigma / \sqrt{N}}$	$z = \dfrac{27.78}{20 / \sqrt{100}} = 13.9$

COMPARING A SAMPLE MEAN TO A POPULATION WITH KNOWN MEAN AND UNKNOWN VARIANCE – THE ONE SAMPLE *t*-TEST

Terms you should learn:
Critical ratio
t-statistic
Degrees of freedom
t-distribution

Concepts you should master:
What the null hypothesis really says
Critical ratio as a generalized distance
The uncertainty due to estimation and the price in degrees of freedom

Case study (continued)

Your original research question was whether you could lower cholesterol below the levels found in the normal population levels using only rest at a holiday resort, good food and plenty of exercise. In the last section we found out that you could. There is, however, still one point left to be settled. Is there really a significant difference between the mean serum cholesterol level of the treated population and the 'danger' level (200 mg/dl). Was the serum cholesterol level in your population lowered far enough to allow you to conclude that the treatment really induces good cardiovascular health?

Simple inspection shows that 209 mg/dl is greater than 200 mg/dl. The statistical question is, 'How likely is it that the serum levels I observed could have been drawn from a population which had a true mean of 200 mg/dl?'

The data

The data are the same as those given above. Cholesterol levels after treatment had a mean of 209.22 mg/dl and a standard deviation of 16.34 mg/dl. Are these levels statistically indistinguishable from those seen in a population with a mean at the 'danger' level of 200 mg/dl?

The data can be plotted in a histogram like that shown in Figure 16. In this instance we do not know the standard deviation of the cholesterol levels in the 'healthy' population, only that of the sample drawn from the target group. The 'danger level' is marked on the graph. The question now is whether a distribution like that shown in Figure 16 would result if the sample were drawn from a population with the true mean of 200 mg/dl. Although the difference between the sample mean and the standard does not appear to be 'far enough' away to enable the null hypothesis to be rejected, you must still translate the difference in means into noise units (and then into a probability measure) to be absolutely sure. The issue of experimental design, power, and your ability to detect small differences is discussed in the chapter on Experimental Design. For the time being, you would be satisfied to merely claim that there is a 'significant' decrease in mean serum cholesterol levels from those of the normal population after the holiday regimen, and that the levels achieved are indistinguishable from those of a healthy population.

Data analysis

The cut-off point that you would like your population to attain is an average of 200 mg/dl, but a higher mean serum cholesterol would be acceptable if it was statistically indistinguishable from this value. The statistical question becomes 'What is the likelihood that, after treatment, the experimentally measured serum cholesterol level of 209.22 mg/dl could have been drawn from a population which has a mean serum cholesterol of 200 mg/dl?' This question forms the basis for the null hypothesis.

Calculation of this likelihood using a standard analysis produces the result shown in Table 6.

Table 6

	After treatment
Mean	209.22
SD	16.34
t-statistic	5.61
p-value	< 0.001
Degrees of freedom	99
Critical value (1-sided)	1.67
Critical value (2-sided)	1.99

You can decide whether your treatment is really as good as you think it is by using only the information in this table. We will show you how these numbers were calculated below, but for the time being, let us look at what these results are actually saying. The 'after treatment' mean serum cholesterol level is 5.61 noise units away from the 'healthy' population mean of 200 mg/dl (the t-statistic). The critical value for this study fixes the Type I error rate to 5%. Since you only want to test the differences when the test group has a serum cholesterol level of more than 200 mg/dl, your critical value is 1.67. For the sake of completeness we are including the two-sided critical value (1.99) in case you wanted to determine the ability to lower cholesterol levels to significantly below 200 mg/dl.

The question and the experiment

The original study began with a question about the efficacy of a holiday regimen as a form of cardiovascular therapy. You showed that this intervention lowered serum cholesterol levels to below those of the general population, and your new concern is whether you have removed your patients from the 'danger zone'. The experiment is now asking is whether **on average**, serum cholesterol levels after treatment are indistinguishable from a population ('healthy') which has a mean level of 200 mg/dl.

Assumptions

The scientific assumptions underlying the use of the one sample t-test in this experiment are:

(1) the true mean of an underlying population (in this case 'healthy' people) is known;
(2) the true standard deviation of that underlying population is not known;
(3) scientifically valid data are being measured;
(4) the underlying population is truly homogeneous;
(5) you only care if the mean of your population is statistically significantly greater than that of the 'healthy' population.

The statistical assumptions required for this test are:

(1) the measures are continuous (e.g. not all-or-none effects, scores, etc.);
(2) the measures are drawn from an underlying distribution which is normal;
(3) the sample population is chosen randomly from the underlying population;

(4) there are enough subjects in the sample to detect small differences, or at least differences that can carry with them biological consequences (see Experiment Design, below, for a discussion of statistical power).

The test – the equation, what it means, and the distance it measures

The null hypothesis is the same as that given for the one sample *z*-test above:

$$H_0: \mu = \mu_0 \quad \text{or} \quad H_0: \mu - \mu_0 = 0$$

It is assumed that the mean of a sample population is the same as the mean of the underlying population. In the analysis you are going to ask whether your observation is consistent with that assumption; you must therefore calculate the chances of seeing differences between the sample mean and the true mean as large as you did, assuming the null hypothesis is true.

The equation which tests this assumption is given by:

$$t = \frac{\overline{X} - \mu_0}{s/\sqrt{N}}$$

where $\overline{X}$, s, and N represent the sample mean, sample standard deviation and sample size, and μ_0 represents the true mean (200 mg/dl, i.e. your 'danger level') of the underlying population.

This equation says almost the same thing, statistically, that the critical ratio says for the one-sample *z*-test. We are still interested in a difference between a sample mean (a random variable) and a fixed point. The equation is normalized using the only measure we have of noise, the standard deviation of the sample. In fact, in some texts, this equation is also called the critical ratio: by giving it the same name from test to test, we are implicitly confirming that all inferential statistics are really nothing more than distance measurements associated with probability measures.

Again, the numerator is a difference, Δ, which is still a random variable. This test asks whether the mean of the Δ distribution is far enough away from the true mean (200 mg/dl) to cause us concern. Again we assume that $\mu_0 = \overline{X}$, i.e. we assume the null hypothesis is true, and that the sample is drawn from an underlying population for which the mean level is 200 mg/dl. The Δ distribution should, therefore, have a mean of zero, and the test is thus actually asking how far away from zero your particular Δ (9.22 mg/dl) must be before you begin to question the wisdom of assuming that the null hypothesis is true.

We do not, however, know the true standard deviation of the population, σ, as we did for the one-sample z-test: we have only an estimate of σ in s. Since we are again asking about differences in **mean** serum cholesterol levels, we need to use a standard error-like normalizing factor in the denominator of our ratio. Replacing σ by s and proceeding as above we are almost home.

This critical ratio is a random variable, and is represented by its own distribution, which has a mean of zero. It is, however, different from the standard normal curve used for the z-test because we do not know the value of σ. We have to pay a price in uncertainty for estimating σ by s. The new distribution, known as the *t-distribution*, takes into account this uncertainty by accounting for the precision of our estimate of σ.

Intuitively, increased uncertainty in the estimate of σ suggests that the two means should be farther apart (in noise units) to achieve the same 'comfort level' in rejecting the null hypothesis. In other words, once you have fixed your Type I error rate at 5%, for example, increasing uncertainty in σ requires the distance between $\overline{X}$ and μ_0 to be greater than that required for the z-test. It would be expected that as fewer points are taken in the sample, the estimate of σ should become less precise. The shape of the t-distribution should therefore be a function of the sample size, forcing more and more probability measures into the tail as N becomes smaller. This dependence of the t-distribution on the sample size harks back to the earlier reference to the *degrees of freedom* of the data set (see Descriptive Statistics). Degrees of freedom is the number of independent values sampled minus the number of estimates made from them. In essence, they act as information coins: you have to pay for the right to estimate parameters, such as the standard deviation, from data. Mathematically, the argument is that if you have N data points and you want to estimate the mean, you can let N–1 of them float anywhere you want. However, once you have fixed those N–1 values (i.e. when you have randomly chosen your N–1 sample points), the Nth value is completely determined by your estimate of the mean. Since it is determined, it is no longer a random variable you are sampling, and the number of independent sample points is actually reduced by one.

In practice, you are really not fixing a mean and then sampling your statistical population to fit it. Mathematically, however, once you have N sample points, you are estimating the true mean, μ, of your underlying population by $\overline{X}$. Without even knowing it, you are then asking about all the possible samples that you could have taken, given $\overline{X} = \mu$. From that implicit question the limit of $N-1$ degrees of freedom is determined.

The same situation arises for the *t*-distribution. You can only freely sample $N-1$ points before the last is strictly determined; there are, therefore only $N-1$ degrees of freedom in your sample. The corollary to all this mathematics is that the standard normal distribution is just a *t*-distribution with infinite degrees of freedom: as you sample more and more data points from a statistical population, the estimate of σ by s becomes so precise that, in the end, you know the value of σ.

The dependence of the *t*-distribution on the degrees of freedom is shown graphically in Figure 18, which shows three curves drawn as overlays of the standard normal distribution. The critical values of each are also marked, and it can be seen that these move farther away from zero as the number of degrees of freedom decreases. This corroborates mathematically our intuitive notion that the less we know about our statistical population, i.e. the smaller the sample, the greater the distance between means must be for us to maintain a fixed error rate.

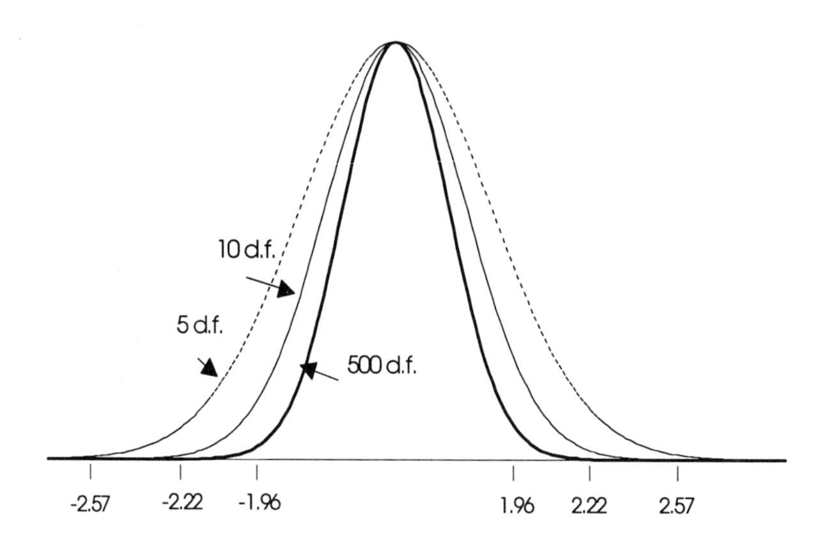

Figure 18 Dependence of the *t*-distribution on degrees of freedom

Returning to our example

The calculations associated with our example data are outlined in Table 7.

Table 7

Formula	Example
$\Delta = \overline{X} - \mu_0$	$\Delta = 209.22 - 200$ $= 9.22$
σ is estimated by s	$s = 16.34$
$t = \dfrac{\overline{X} - \mu_0}{s/\sqrt{N-1}}$	$t = \dfrac{9.22}{16.34/\sqrt{100}} = 5.61$
Degrees of freedom = N–1	Degrees of freedom = 99

We test the null hypothesis as follows. First, calculate the numerator for the critical ratio, i.e. the difference (9.22 mg/dl). Then calculate the denominator (based on s and N), which in this case is 1.63 mg/dl. The distance between the sample mean serum cholesterol level and 200 mg/dl is 5.61 noise units. To determine the probability of observing a sample mean that is 5.61 or more noise units away from the standard, given that the null hypothesis is true (i.e. to link the distance measure to a probability), we turn to the t-distribution with 99 degrees of freedom, and find that the likelihood of observing a difference this big is less than 0.001.

Based on these data, the chance of seeing a sample mean 'this much' greater than 200, given the null hypothesis is true, is much less than 5%, and we must conclude that while the 'holiday therapy' decreased serum cholesterol levels significantly when compared with those of the general population, it did not lower the levels far enough, at least after three months, to enable the treated patients to be declared 'healthy'.

COMPARING BEFORE AND AFTER DATA – THE TWO SAMPLE PAIRED *t*-TEST

Terms you should learn:

Critical ratio
t-statistic
Degrees of freedom
t-distribution

Concepts you should master:

What the null hypothesis really says
Critical ratio as a generalized distance
The uncertainty due to estimation and the price in degrees of freedom

Case study (continued)

Your treatment regimen lowered serum cholesterol levels to below those of the ordinary population, but these levels were not low enough to declare your patients out of the danger zone, i.e. the mean level in the treated population was significantly above 200 mg/dl. The question then arose of whether other risk factors had been beneficially altered by the treatment. A literature search showed that for non-smoking males over 52 years old, overall body weight was a second significant risk factor for measuring cardiovascular health. Therefore, if that group alone showed a significant decrease in body weight after treatment, you might still claim that the therapy was beneficial to good cardiovascular health in some people.

Review of the records identified 31 men aged 52 or more in the sample. The medical records included their weights at the start and at the end of the treatment, enabling you to determine whether the treatment produced significant weight loss in this subpopulation.

The data

The body weights of these individuals before and after therapy are given in Table 8. The mean weight loss was 13.35 pounds (standard deviation 9.92 pounds).

Data analysis

You would like to show that, **on average**, these men lost weight after 3 months of treatment in a holiday resort. The statistical question becomes 'What is the likelihood that a mean weight loss of 13.35 pounds would be observed if there was really no effect of therapy?'

Calculating this likelihood using a standard analysis produces the results shown in Table 9.

Table 8

Patient	Weight before	Weight after	Change
1	204	206	+2
2	223	204	−19
3	214	221	+7
4	229	216	−13
.	.	.	.
.	.	.	.
.	.	.	.
30	237	209	−28
31	223	207	−16
Mean	227.31	213.96	−13.35
SD	14.12	12.88	9.92

Table 9

	After treatment
Mean	−13.35
SD	9.92
t-statistic	− 7.49
p-value	< 0.001
Degrees of freedom	30
Critical value (1-sided)	1.70
Critical value (2-sided)	2.04

After treatment, mean weight loss is 7.49 noise units away from zero (what you would expect if your treatment had no effect at all). If the Type I error rate is fixed at 5%, then the critical value for this study is 1.70. If you had been interested in whether there had been a weight change of any kind, and not just a loss, you would have had to perform the test using a two-sided test, and the critical value would be 2.04. Recall from the last section that when there were 99 degrees of freedom in the analysis our critical values were slightly smaller. In this case, the degrees of freedom are 30 (31 pairs of data – 1 for estimation). The estimate of σ is therefore less precise (31 patients vs. 100).

The question and your experiment

The original study looked at the efficacy of a holiday resort as a form of cardiovascular therapy. This intervention lowered serum cholesterol levels to below those of the general population, but the treated population was still not 'healthy'. Since the combination of lower serum cholesterol level and significant weight loss in men over 52 is probably a good predictor of increased cardiovascular health, you are now asking whether the treatment makes 52-year-old men more healthy. What the experiment is asking is, 'on average, are the weight losses observed significantly more than nothing?'

Assumptions

The scientific assumptions underlying the two-sample paired *t*-test are:

(1) you must estimate the true mean difference between pairs for an underlying population;
(2) you must estimate the true standard deviation of the distribution of differences for that underlying population;
(3) the underlying population is truly homogeneous.

The statistical assumptions required for this test are:

(1) that your measures are continuous (e.g. not all-or-none effects, scores, etc.);
(2) that they are drawn from an underlying distribution which is normal;
(3) the sample population is chosen randomly from the underlying population;
(4) that you have true pairing of your data.

The test – the equation, what it means, and the distance it measures

The null hypothesis is a variation on a theme: are the mean differences observed within my pairs different from zero, i.e. could differences this big have arisen in a sample drawn from a population with a mean of zero and a standard deviation of *s*? In both the one-sample *z*-test and one-sample *t*-test, we asked whether our sample mean was different from a fixed point. In this case the fixed point is zero. The null hypothesis therefore simplifies to:

$$H_0: \mu_D = 0$$

In English, you are assuming that there is no post-therapy weight loss, i.e. the mean of the underlying population is actually zero. In the analysis, you are going to ask whether the mean of the differences observed **within each subject** is consistent with that assumption. This requires calculation of the likelihood of observing a sample mean as large as that seen when the null hypothesis is true.

The equation which tests this assumption is given by:

$$t = \frac{\overline{D} - 0}{s_D / \sqrt{N}}$$

where D, s_D and N represent the sample mean change, sample standard deviation of the changes, and the number of pairs (the true sample size in this experiment) and zero is the hypothesized weight loss.

When μ is replaced by zero in the one-sample t-test, the equations are identical. We are still interested in a difference between a sample mean (a random variable) and a fixed point, and the equation is still being normalized using the only measure we have of noise, the standard deviation of the sample. We should, therefore, not be too surprised that this equation is also called the critical ratio. Implicitly, the numerator is still the difference, Δ, but this test asks whether the mean of the Δ distribution (which can now be estimated by D) is far enough away from zero and not μ to concern us.

Again, we must estimate σ using s_D. To calculate the distance in noise units, we again rely upon a standard error-like normalizing factor to adjust our mean difference to a noise measure. This critical ratio is still a random variable, and it is still represented by the t-distribution. This should not come as any great surprise, since, so far, the two tests seem to be identical. In fact, all we did was replace μ by zero and change our analysis variable to the mean change within each patient. Once we have the distance in noise units we need only associate that distance with the correct probability measure to assess our risk of rejecting the null hypothesis wrongly. As in the one-sample case, we must modify our probability estimate by the amount of uncertainty in the estimate of σ. This time we use the number of pairs minus one as our index, because we have only 31 (and not 62) **independent** measures of weight difference.

Notice that we did not compare the mean body weights before and after treatment (227 pounds and 213 pounds, respectively) directly, but used the change in weight (13.35 pounds) within each patient. This is an important factor in this type of design: using the patient as his/her own control produces perfect pairing. As you will see in the next section, when you cannot take advantage of this *within-subject* design, you are required to estimate two means and a pooled estimate of the variance to reach the same point.

Why does this work so well, and what do we mean by *paired data*? Consider the following premise. The variance observed within a subject or patient is usually significantly less than that seen across different subjects or patients. The reason is that, usually, the values measured within each subject are correlated and not independent. If each member of the group of men had lost exactly 13.35 pounds, the standard deviation of the 'after treatment' weights would have been identical to that of the 'before treatment' weights (14.12 pounds), but the standard deviation of the differences would have been zero. It is this lack of independence that makes the data paired. Twins can be used as each other's control, making a paired *t*-test possible. Left and right sides of the body in the same subject can be used in the same way. Subjects matched for age and sex can sometimes be considered paired if the outcome measure can be shown to depend only upon treatment and not on matching factors. That is a very large assumption, however, and it is more conservative not to assume that the age-sex matched subjects are pairs, but to consider the two populations as independent. That case is considered below – but this special case of the one-sample *t*-test can be performed in any case where the subjects can be matched or paired.

Returning to our example

The calculations associated with our example data are outlined in Table 10. We can calculate the numerator for the critical ratio, i.e. the difference, which is –13.35 pounds, and the denominator (based on s_D and N), 1.78 pounds. –13.35 pounds is –7.49 noise units away from zero. To determine the probability of observing a mean 7.49 or more noise units away from zero when the null hypothesis is true, we turn to the *t*-distribution with 30 degrees of freedom, and see that the likelihood of observing a difference this big is less than 0.001.

Table 10

Formula	Example
$\Delta = \overline{D}(-0)$	$\Delta = -13.35$
σ is estimated by s_D	$s_D = 9.92$
$t = \dfrac{\overline{D}}{s_D / \sqrt{N}}$	$t = \dfrac{-13.35}{9.92 / \sqrt{31}} = -7.49$
Degrees of freedom = number of pairs minus 1	Degrees of freedom = 30

Based on these data, the probability of wrongly declaring the null hypothesis false is much less than 5%. We can, therefore, conclude that you were able to decrease body weight in 52-year-old men significantly by sending them to a holiday resort and forcing them to eat healthily and exercise regularly.

COMPARING TWO MEANS – THE TWO SAMPLE UNPAIRED *t*-TEST

Terms you should learn:

Critical ratio
Pooled variance

Concepts you should master:

What the null hypothesis really says
Critical ratio as a distance
The true meaning of the *p*-value in this test

Case study

Suppose you know that antagonizing a specific receptor in the smooth muscle of the ureter will increase urine output, and suppose you believe that a certain compound can be a potent antagonist of that receptor. The question you have is 'If I administer this compound to dogs, will urine output increase?'

You have developed a study plan using two groups of dogs: one receiving an i.m. injection of saline and the other receiving an injection of your test compound (0.3 mg/kg). Each dog is catheterized, assigned to a treatment arm, and urine is collected for 3 hours after dosing and measured in milliliters. Both groups receive water *ad lib* for 24 hours after surgical preparation.

Preliminary data on urine output in dogs suggest that a 40 ml increase in urine output is biologically meaningful, and you have determined that six dogs in each group is sufficient to detect this increase. The 12 dogs were randomized into the groups so that their assignment to saline or drug treatment was completely unbiased. We have already discussed why randomization is basic to the theory of descriptive statistics: the same arguments apply here.

Details regarding the experiment design phase of this project, including issues relating to power, randomization, etc. are presented later in our discussion of Experimental Design. We believe that good experiment design begins as early as the animal handling phase of the study. Anything that can be done to account for selection factors, bias or noise will help us eventually to separate the true biological signal from naturally occurring variation between subjects.

The data

A difference of 44.2 ml of urine (203.5 –159.3) was observed between the drug- and saline-treated dogs in the experiment. The data collected from your experiment, along with preliminary summary statistics, are presented in Table 11. The dot plot obtained from these data was used as an example in Chapter 1 (Figure 13) and is reproduced here for easy reference.

Table 11

	Saline	Drug
	200	242
	179	225
	167	222
	152	195
	145	178
	113	159
Mean	159.3	203.5
SD	30.0	31.6

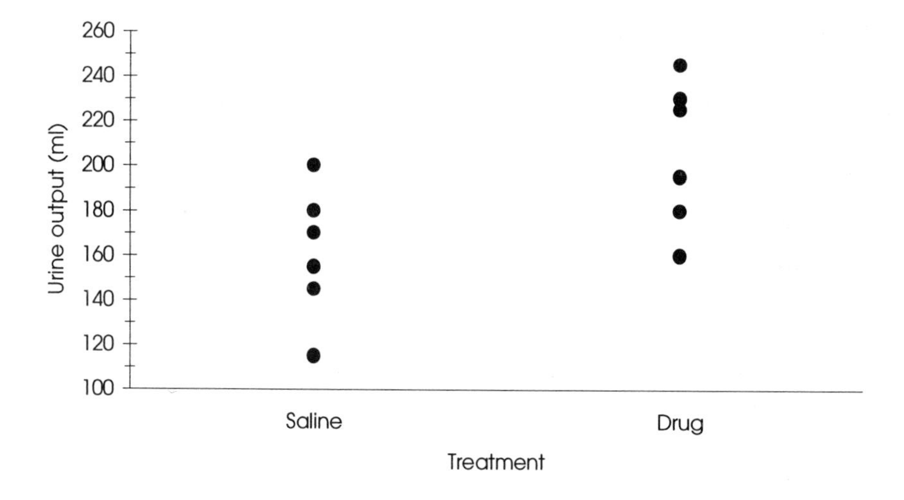

Figure 13 Urine output in dogs treated with saline or drug

Data analysis

We know that the observed difference in sample means, 44.2 ml, is actually a random 'snapshot' drawn from the distribution of all differences in urine output for drug-treated dogs compared with saline-treated animals. The measure 'difference', in fact, is a random variable, and is thus subject to variability in its own distribution. In another experiment we might have observed a difference as high as 60.0 ml or maybe even as low as 30.0 ml. The statistical question becomes 'Is the experimentally measured difference of 44.2 ml so large that it is unlikely to have been drawn from a population with a mean of zero?' This question forms the basis for the null hypothesis. We have calculated this chance using a common spreadsheet program (Table 12). Most spreadsheet and data analysis software packages can perform this calculation. In this case we have used EXCEL (Analysis Tools, *t*-test, Two-sample assuming equal variances).

Table 12 *t*-Test: two-sample, assuming equal variances

	Variable 1		Variable 2
Mean	203.5		159.3
Variance	997.9		901.1
Observations	6		6
Pooled variance		949.5	
d.f.		10	
t		2.48	
$P(T \leq t)$ one-tail		0.016	
$t_{Critical}$ one-tail		1.812	
$P(T \leq t)$ two-tail		0.03	
$t_{Critical}$ two-tail		2.23	

The software produces a table which provides all the information needed to make a decision about the efficacy of the test compound. Later on we will show you how these numbers were calculated, but for the time being we can trust EXCEL to calculate the *t*-test statistic as 2.48 noise units. The computer provides you with both resources you need to evaluate your results:

(1) the probability of observing your difference (2.48 noise units) given the null hypothesis is true, i.e. the *p*-value (labeled $P(T \leq t)$);
(2) the critical value for the *t*-statistic, the measure of the least difference in noise units acceptable for a preordained risk level for

error (e.g. in this case it was predetermined to be at a 5% Type I error rate and equal to 2.23 noise units for a two-tail test).

Suppose we expect an increase in urine output due to drug treatment and a one-tail test is warranted. The probability associated with the *t*-test statistic (the *p*-value) and the point at which a 5% error rate will occur (the critical value) are shown schematically in Figure 19.

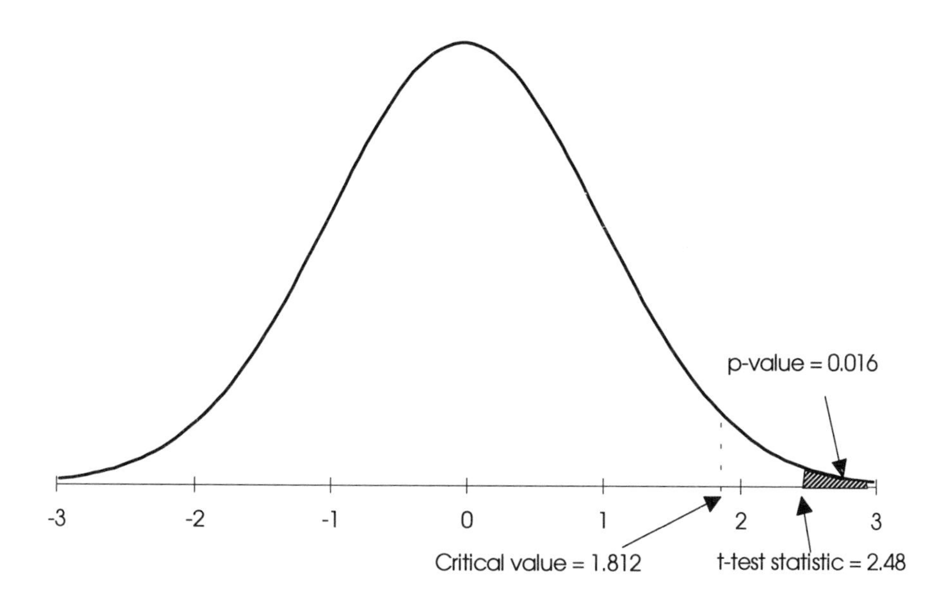

Figure 19

The 'bell-shaped' curve depicts the *t*-probability distribution with 10 degrees of freedom. Remember, we saw this earlier in our one-sample tests. In this case, the curve represents the distribution of all possible differences we would expect to see if, in fact, the drug did not increase urine output. The *t*-statistic, or the distance measured in noise units, falls in the extreme end, or tail, of this distribution, suggesting that the likelihood of seeing a difference this big (44.2 ml of urine) given the null hypothesis is true (that is, that your samples were drawn from a population centered with mean zero) is very small. We usually decide that a result is 'significant' if that likelihood is 'small enough' (beyond the critical value). In this example, the critical value was preset to limit our Type I errors to only 5%. The t-statistic falls outside this range, and we say that there is a 'significant' increase in urine flow.

In the following parts of this section we will discuss the details of this study, and then return to these data to show you how EXCEL actually performs the calculations.

The question and your experiment

Every experiment starts with a question and a possible solution. A valid assay is then established to test the hypothesis: the outcome of your test can be measured (e.g. increased urine flow) and the experimental model allows results to be obtained in a fairly 'reasonable' way – the measures are consistent and continuous. What you are asking is 'Does this test compound really antagonize the receptor in the smooth muscle of the ureter?' What your experiment is asking is 'on average, do treated dogs yield an increased urine flow compared with controls?'

Assumptions

The scientific assumptions underlying the two-sample t-test are:

(1) you are comparing only two groups;
(2) they have both experienced comparable experimental conditions;
(3) the treatment group differs from the controls only by the planned intervention;
(4) the measurements taken reflect the true effect being looked for, i.e. the effects measured are not spurious side effects of the experiment.

The statistical assumptions required for this test are:

(1) the measures are continuous (e.g. not all-or-none effects, scores, etc.);
(2) they are drawn from underlying distributions which are normal;
(3) they have equal variance (although there is a method around which does not make this assumption).

The test – the equation, what it means, and the distance it measures

In general, the null hypothesis is given by:

$$H_0: \mu_1 = \mu_2 \quad \text{or} \quad H_0: \mu_1 - \mu_2 = 0$$

In English, this says that the two underlying populations have the same mean, and you are going to test whether the observations made in the experimental samples drawn from that population are consistent with that assumption. In other words 'What are the chances of observing differences between two sample means 'this big' given they were drawn from one underlying population?' The statistical assumptions upon which this test is based are actually restating part of this null hypothesis.

The equation which tests this assumption is given by:

$$t = \frac{\overline{X}_1 - \overline{X}_2}{s_p \sqrt{\frac{1}{N_1} + \frac{1}{N_2}}}$$

where

$$s_p^2 = \frac{(N_1 - 1)s_2^2 + (N_2 - 1)s_2^2}{N_1 + N_2 - 2}$$

and the $\overline{X}$'s, N's and s^2's represent the sample means, sample sizes, and sample variances of each of the two populations respectively.

What do these two equations tell you? The first equation, still known as *the critical ratio*, says that we are interested in a difference of means $(\overline{X}_1 - \overline{X}_2)$ which is normalized in some way for the noise of the measures. If we think about $\overline{X}_1 - \overline{X}_2$ as a single variable the test asks about a single realization drawn from the distribution of the random variable, **difference in means**. Under the null hypothesis, i.e. assuming $\mu_1 = \mu_2$, this distribution should have a mean of zero. You know, however, that the difference between two random sample means will not be exactly zero, so how far away from zero must it be before you begin to believe that you may be mistaken in assuming the null hypothesis?

To answer that question we need to know the spread around zero expected for $\overline{X}_1 - \overline{X}_2$ when the null hypothesis is true. Given this, we are then able determine in a probability metric how far from zero D $(= \overline{X}_1 - \overline{X}_2)$ really is. That noise measure, based on s_p^2, is *the pooled variance* of the two populations and is supplied in the second equation. It is an estimate of the variance, and is a weighted average of the two sample variances s_1^2 and s_2^2. The weights are given by the ratios of the two sample sizes (N_1 and N_2), to the total observations taken: that is easier to see as

$$s_p^2 = (\text{weight}_1)s_1^2 + (\text{weight}_2)s_2^2$$

$$= [\frac{(N_1 - 1)}{N_1 + N_2 - 2}]s_1^2 + [\frac{(N_2 - 1)}{N_1 + N_2 - 2}]s_2^2$$

What we want to know is how likely we are to see a sample difference as large or larger than $(\overline{X}_1 - \overline{X}_2)$ drawn from a population with mean zero and variance s_p^2. Remember that the standard deviation of a distribution of sample means is the standard error of the mean, and that it is calculated by dividing the standard deviation of the sample by the square root of N, the sample size. We happen to be looking at the distribution of the difference of two means, the random variable, D. The theoretical variance of the difference is given as:

$$\sigma^2_{\overline{X}_1 - \overline{X}_2} = \frac{\sigma^2_{X_1}}{N_1} + \frac{\sigma^2_{X_2}}{N_2}$$

The standard deviation of the distribution of mean differences is then given by:

$$\sigma_{\overline{X}_1 - \overline{X}_2} = \sqrt{\frac{\sigma^2_{X_1}}{N_1} + \frac{\sigma^2_{X_2}}{N_2}}$$

We are, however, assuming that the values of $\overline{X}$ are drawn from the same underlying distribution. Remember the null hypothesis? In theory at least, the values of σ_X^2 should both equal σ^2, the true underlying variance, and s_p^2 is our best estimate for that. Therefore, substituting s_p^2 into this equation and factoring it through the radical yields:

$$\sigma_{\overline{X}_1-\overline{X}_2} \approx s_p\sqrt{\frac{1}{N_1}+\frac{1}{N_2}}$$

the denominator of the first equation.

Given these calculations, we can now associate the distance calculated in noise units (the difference divided by the noise) with a probability distribution which looks like that shown in Figure 19. These theoretical curves are specific for the distribution called the *t-distribution,* and have been around for about 100 years. They have been tabulated, and were used by EXCEL in our earlier calculation. The degrees of freedom for this test are $(N_1 + N_2 - 2)$: two information coins were spent to estimate the means, $\overline{X}_1$ and $\overline{X}_2$.

Returning to our example

The calculations associated with our example data are outlined in Table 13. The null hypothesis assumes that the two sample populations (treated and control) were drawn from a single underlying population, and hence, have the same mean. The statistical question 'What is the likelihood of observing a difference this big (44.2 ml; 203.5 – 159.3) given the null is true?' associates that difference with a probability measure.

The test is run as follows. We have the numerator for the test (the difference, 44.2 ml) and we calculate s_p^2 as 949.56, using the second equation (since the two sample sizes are the same, it is not too surprising that this is just the average of the two variances). s_p the square root of this value, is therefore 30.81. The distance metric requires we take the square root of 1/6 plus 1/6 to find the standard error (the standard deviation of the distribution of differences of sample means), so that the denominator of the first equation becomes 17.79. This tells us that 1 standard error unit in the distribution of the difference is 17.79 ml of urine, and that 203.50 is 2.48 of those units away from 159.3. Alternatively, we could ask how likely it is that the number 44.2 would be drawn at random from a distribution which has a mean of zero and standard deviation of 17.79. To determine the probability that this actually occurred by chance alone, i.e. to link the distance measure to a probability, the critical values associated with 10 degrees of freedom (6 + 6 – 2) are looked up in the *t*-table. These values are shown in Table 14.

Table 13

Formula	Example
$\overline{X}_1 - \overline{X}_2$	203.5 – 159.3 = 44.2
$s_p^2 = \frac{(N_1-1)s_1^2 + (N_2-1)s_2^2}{N_1+N_2-2}$	$s_p^2 = \frac{(6-1)\bullet 901.1 + (6-1)\bullet 997.9}{6+6-2}$ $= 949.5$
$t = \frac{\overline{X}_1 - \overline{X}_2}{\sqrt{s_p^2(\frac{1}{N_1}+\frac{1}{N_2})}}$	$t = \frac{44.2}{\sqrt{949.5 \bullet (\frac{1}{6}+\frac{1}{6})}} = 2.48$
$t_{\text{Critical}} = t_{N_1+N_2-2}$ One sided t-test	$t_{\text{Critical}} = t_{6+6-2} = 1.812$ One sided t-test

Table 14

One-sided critical values	0.1	0.05	0.02	0.01
10 df	1.812	2.228	2.764	3.169

Based on these data, this result says that the probability of observing a difference as large as 44.2 ml given the null hypothesis is true is less than 5% but more than 2% ($2.228 < 2.48 < 2.764$) The p-value is less than the critical value of 0.05, so we can reject the null hypothesis and be wrong less than one time in 20. This is exactly the result produced by the EXCEL software.

COMPARING THREE OR MORE MEANS – THE ONE WAY ANALYSIS OF VARIANCE

Terms you should learn:

Homoscedastic
Heteroscedastic
ANOVA table – the F-statistic, degrees of freedom and the p-value
Experiment-wise error rate
Pairwise error rate

Concepts you should master:

What the null hypothesis really says
Why comparing variances lets you test means
Homoscedasticity vs. heteroscedasticity – robustness
of the assumptions
Multiple comparisons testing – experiment-wise error vs. by-test error

Foreword

This section introduces a tool with which to compare three or more means across different groups of subjects. Before getting further into these fairly complex calculations, consider why all the means within one analysis need to be compared simultaneously when performing all the pairwise t-tests separately would probably suffice. The reason is that when you decide to perform a t-test, you must choose, *a priori*, the comfort level for erroneously rejecting the null hypothesis. That is equivalent to saying that you are willing to declare two population means different when in fact they are not. Previously, we defined that error rate as the Type I error which, by convention, is usually preset to 5%. If you have a 1 in 20 chance of making a mistake on one trial, however, the composite chance of making that same mistake in two independent trials is clearly no longer 1 in 20, and it is certainly no lower. If you undertook an enormous number of such tests (perhaps infinite) you are guaranteed to make that same mistake, no matter how small each individual probability is. The overall error rate, called the *experiment-wise error rate*, is therefore what you really want to control at the outset of an experiment. The individual trials are tested at the *pairwise error rate*.

As an example of error inflation, suppose the pairwise error rate was set to 5%. While the chance of erroneously rejecting the null hypothesis for one pairwise comparison is exactly 5%, the chance of rejecting the null hypothesis for two pairwise comparisons (e.g. one control group and two treatment arms) is given by:

$$\text{Experiment-wise error} = 1 - (1.0 - 0.05)^2 = 0.0975$$

If an experiment required five pairwise comparisons, the likelihood of erroneously rejecting one or more of the null hypotheses would be 22.6%. In other words, the chance of making a mistake when using the individual tests is approximately 1 in 5 – more than four times the overall comfort level.

Case study

Suppose an experiment tests more than a single pair of treatment arms: for example, a study of the dose–response characteristics of a particular drug. This might involve testing four or five different (usually increasing) doses or concentrations in an assay which is known to be predictive of drug efficacy. The results of this bioassay might be compared with those of a standard reference compound or an untreated control, or perhaps both. Another example might involve screening drugs. This would probably require a number of them to be tested simultaneously. Again you would want to know which of these drugs 'works', and which is better than another. These two experiments are almost identical when it comes to their logistics, but the designs are subtly different. These differences, subtle though they may be, are reflected in the analysis tools applied to the interpretation of the results.

Let us work through the following example. A protein chemist is convinced that a series of molecules she has developed, all based on the same basic protein class, will decrease the proliferation rate of autoreactive T-cells in at least some autoimmune diseases. She is sure that the key to T-cell proliferation is an enzyme which has an active site that is 9 Å wide by 12 Å high by 4 Å deep, and this pocket has a positive charge on its uppermost face. Your laboratory has been given the task of determining if her conjecture is correct. An extensive literature search alerts you to an in vivo proliferation assay in which T-cells are excited by a non-specific adjuvant (a substance which induces immunoreactive cells to multiply) and which reasonably mimics autoimmunity. You have designed your experiment to include the following treatment arms:

(1) T-cells from animals receiving no adjuvant stimulation
(2) T-cells from animals receiving the adjuvant stimulus, but no

drug of any kind.

(3) T-cells from animals receiving the adjuvant and one dose of test compound 1

(4–7) As Group 3 except that they each receive test compounds 2–5 respectively.

Your questions are as follows:

(1) Did the adjuvant work?
(2) Did any of the test compounds actively inhibit the proliferation rate of the T-cells in vivo?
(3) If so, which compounds were they?

In this experiment, T-cells from the spleens of test animals were extracted one day after administration of the adjuvant and stained to indicate DNA replication levels. Those showing twice the normal DNA content, i.e. those in M-phase, were counted in a flow cytometer. The theory is that the number of cells in M-phase is a reasonable marker of the proliferation rate, i.e. is proportional to the total number of cells in the active phase of the cell cycle.

The data

The data obtained from the experiment are given below. The recordings are counts and are presented as a table (Table 15; means and SD) and as a dot plot (Figure 20).

Table 15

Treatment	Counts (mean ± SD)
1. No Adjuvant	109.47 ± 43.76
2. Adjuvant; no drug	1763.28 ± 73.45
3. Adjuvant; test compound 1	1719.81± 77.36
4. Adjuvant; test compound 2	1797.00 ± 99.23
5. Adjuvant; test compound 3	1568.04 ± 86.27
6. Adjuvant; test compound 4	1488.28 ± 97.78
7. Adjuvant; test compound 5	992.75 ± 61.87

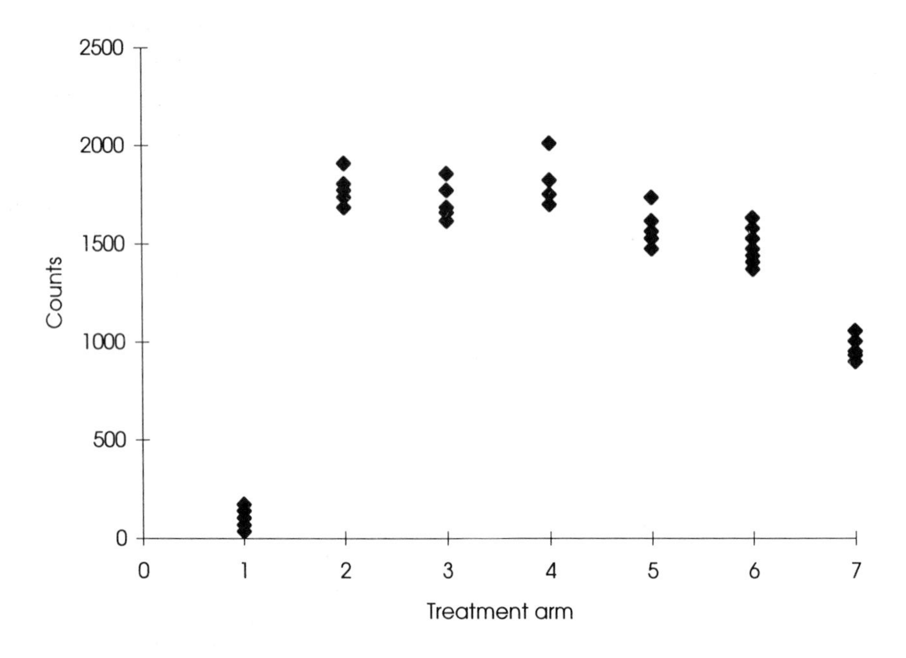

Figure 20 Dot plot of T cell counts: data from Table 15

Data analysis

The analysis of these data is shown in Tables 16–18.

Table 16

	No Adjuvant		Adjuvant No Drug
Mean	109.47		1763.282
SD	43.76		73.45
Observations	7		7
$t_{Critical}$ (12 df)		2.179	
t-statistic		51.2	
p-value		<0.001	

work?', is answered using only these two groups. If you were to include any of the other groups receiving adjuvant, i.e. those also receiving therapy, you would confound the issue with the treatment effects of any active compounds. If we ignore the controls not given adjuvant we can now answer the second question, 'Did any of these compounds significantly inhibit T-cell proliferation after adjuvant therapy?' Note: Sometimes a statistician would like to answer these two questions simultaneously, and there are ways to do that. But for illustrative purposes we have pulled them apart. Table 17, called an *ANOVA table,* is a little more complex than those presented previously, but it provides all the information you need to determine the answer.

Table 17

Source	Sum of squares	Degrees of freedom	Mean square	F-statistic	p-value
Treatment	3148980	5	629796	89.85	<0.0001
Error	252337	36	7009	$F_{\text{Critical}} = 2.48$ $F = 89.95$ Degrees of freedom = (5,36)	
Total	3401317	41			

We will discuss the meaning of each of the entries in Table 18 below, but for now the p-value that answers your second question is given in the last column of Table 17. This value represents the probability of observing the pattern of sample means actually obtained, given that the null hypothesis is true. The null hypothesis in this case is that all six samples (all adjuvant-treated animals) were drawn from the same underlying population. In other words, if each group was chosen at random from a single underlying population, the chances of observing a pattern like this is less than 1 in 10 000.

Table 18 provides you with the p-values derived using the Analysis of Variance for each individual test comparing adjuvant-stimulated controls (no drug) with drug-treated subjects. These pairwise comparisons are made simultaneously within the context of the overall analysis. The details of how they were performed are discussed below, but this extension of the two-sample t-test allows *experiment-wise error rates* to be controlled for and separated from the corporate error you would encounter if you performed five individual tests each with its own *pairwise error rate.*

Table 18

Treatment	Counts (mean ± SD)
1. Adjuvant; no drug	1763.28 ± 73.45
2. Adjuvant; test compound 1	1719.81 ± 77.36 (ns vs. Group 1)
3. Adjuvant; test compound 2	1797.00 ± 99.23 (ns vs. Group 1)
4. Adjuvant; test compound 3	1568.04 ± 86.27 ($p < 0.05$ vs. Group 1)
5. Adjuvant; test compound 4	1488.28 ± 97.78 ($p < 0.05$ vs. Group 1)
6. Adjuvant; test compound 5	992.75 ± 61.87 ($p < 0.05$ vs. Group 1)

The question and your experiment

The study began with a question about the efficacy of a family of proteins as anti-proliferative agents in some autoimmune diseases and a reasonable in vivo model of T-cell proliferation which you believe is predictive of these autoimmune reactions. Using standard laboratory practices and good experiment design, you decided that counting the number of spleen cells in M-phase one day after treatment would give a reasonable indication of the antiproliferative activity of any protein in the test system. What you are asking is 'Does my model of autoimmunity really work – i.e. do the cells begin to proliferate when animals are challenged with this non-specific adjuvant?' If so, which, if any, of these proteins really inhibits this kind of proliferative growth? What your experiment is asking is, 'on average, are there more cells in M-phase in the spleens of adjuvant-treated animals than in the spleens of animals receiving no adjuvant?, and if so, on average, do the animals in any of the treatment arms have fewer M-phase cell counts than the appropriate control?'

Statistically, you are trying to determine the chances of observing a distribution of sample means as disparate as that observed, given that these samples were all drawn from the same underlying population.

Assumptions

The scientific assumptions underlying the use of the Analysis of Variance are:

(1) the true mean of the underlying population is not known;
(2) the true standard deviation is not known;
(3) the number of cells in M-phase is a scientifically valid measure of autoreactive T-cell proliferation rates;
(4) the underlying populations are truly homogeneous, i.e. only autoreactive T-cells are being measured in the assay.

The statistical assumptions required for this test are:

(1) the measures are continuous (e.g. not all-or-none effects, scores, etc.);
(2) the measures are drawn from underlying distributions which are normal, and have the same mean and variance;
(3) the sample populations are chosen randomly from the underlying populations;
(4) each treatment arm contains enough subjects to detect small differences in the proliferation rates, or at least differences large enough to be biologically relevant.

The assumption of equal variance (assumption 2 above) is known as *homoscedasticity* and is very important (see below). The existence of unequal variances among the populations is called *heteroscedasticity*.

The test – the equation, what it means, and the distance it measures

In general, the null hypothesis is given by:

$$H_0: \mu_1 = \mu_2 = \mu_3 = \mu_4 = \mu_5 = \mu_6 = \mu$$

This equation says that all six underlying populations have the same mean, μ. Under these conditions, the sample means should all be valid estimators of μ, and the only variation about the true mean should be due to variations between samples. The ANOVA tool tests whether the sample means observed are consistent with that assumption. In other words, 'What are the chances of observing a pattern of mean responses (i.e. variations) 'this big', given that all the samples were drawn from the same underlying population?'

Suppose all the samples were drawn from a single underlying population. Then one would expect the mean of the sample variances and the variance of the sample means to be related somehow, and to provide identical estimates of the underlying population variance, σ^2. This is why we must assume *homoscedasticity*.

By now, you should know that the sample means will not all be the same. If they were you would have serious doubts about the validity of your assay. You should therefore expect some variation in your estimates of the true mean. If the null hypothesis is true, however, the sample means should be fairly close to each other, since each actually represents a single random sample from the same underlying population. If the spread of the sample means is too large, then one or more of them is 'sticking out' from the others, and these deviants were probably derived from different underlying populations (i.e. $\mu_i \neq \mu$ for the *i*th group). We therefore need to determine how far apart these sample means must be (i.e. what is their variance?) and to ask if that variance is 'big enough' to indicate that the sample means probably did not come from a single underlying source. The best way to do that is to somehow compare the variance of the sample means with the variance of the true underlying population, σ^2. One estimate of that variance is derived from the mean of the sample variances. The calculations are as follows:

Each treatment arm yields a sample mean and sample variance. In the previous sections we called these statistics $\overline{X}$ and s^2. Recall that the calculation of s^2 looked like :

$$s^2 = \frac{\sum(X_i - \overline{X})^2}{N-1}$$

For the sake of this discussion we have to add a complication to this equation. Each sample has its own variance. Therefore we need to identify each treatment arm with its own subscript. For lack of anything better, we will use the letter '*j*' to indicate that we are discussing treatment group *j*. The equation then becomes:

$$s_j^2 = \frac{\sum(X_{ij} - \overline{X}_j)^2}{N_j - 1}$$

If all the treatment arms have the same number of subjects, then all the N_j are equal, and we can simply calculate the 'average variance' from the samples to derive the first estimate of the underlying variance, σ^2. Now suppose that the number of subjects in each group is not the same. Remember how we were able to calculate an estimate of the underlying variance using the pooled variance in the two-sample *t*-test? The analysis of variance does essentially the same thing. If the experiment has k different groups (in this case $k = 6$), with sample sizes $N_1, N_2, \ldots N_k$, then, in calculations similar to those used for the pooled variance estimate in the two sample *t*-test, the generalized pooled variance estimate has $(N_1-1) + (N_2-1) + \ldots + (N_k-1)$ degrees of freedom.

That estimate for σ^2 is fairly straightforward, but what we really want to know is how disparate our sample means are. That calculation is just as straightforward. Each of the k treatment groups (in this case 6) yields a sample mean, $\overline{X}_j$. The mean of the means, the grand mean, $\overline{\overline{X}}$, is calculated in the usual way. The variance of the means is given by:

$$s^2_{\overline{X}} = \frac{\sum(\overline{X}_i - \overline{\overline{X}})^2}{k - 1}$$

You already know that the standard deviation of the distribution of sample means is really the standard error of the mean, which is the population standard deviation divided by the square root of the sample size. In this case we have k groups each of size N. That is like taking k random points (in this case 6) from the underlying population of the means derived from samples of size N (in this case 7). The second estimate of the true variance, σ^2, is therefore given as:

$$s^2 = N \bullet s^2_{\overline{X}}$$

If one of the means is sticking out 'too far', this is the estimate that would show it: it would be 'too big'.

Recall that all of these machinations are based upon *estimates* of the variances. Then their ratios are random variables and, by the rules about random variables, must be associated with a probability distribution.

This distribution is, in principle, just like those we used for the z-statistic and the various t-statistics, i.e. it associates a probability with a noise metric. This distribution, known as the *F-distribution,* represents the likelihood of observing all possible ratios of variances given that the true ratio should be 1, i.e. the variances are equal. The F-distribution therefore indicates how likely it is to be 'this far away from 1', given that the null hypothesis is true. In that sense, we treat the F-statistic, the test of the ratio, in the exact same way as we treated the t-tests above.

Now for the tricky part. We need to identify the degrees of freedom for each estimate of σ^2. Remember, the degrees of freedom are like information coins: you have a certain amount of information to use, and you have to spend one of your coins for each estimate you make. The estimate of σ^2 using the average of the variances has only Nk independent pieces of information, and one information coin is used to estimate each mean (just as in the t-tests described above). The first method of estimating σ^2 therefore has $(Nk - k)$ degrees of freedom.

The second method of estimating σ^2 actually looks at the variance of the means. What you are really saying is 'I have k samples, each of size N, from which to estimate this variance'– you therefore have $N \times k$ pieces of information. You have to estimate k different means, however, from which you can derive one variance estimate. The number of degrees of freedom for this estimate of σ^2 is therefore $k - 1$.

How can the two be compared? The null hypothesis says that the two estimates of the variance should be the same and that their ratio should be 1. Dividing the second estimate by the first should give a value of 1, or at least near 1. The probability of being 'this far away' from 1 when the null hypothesis is true is derived from tables of the F-distribution, which depends upon both estimates of σ^2. This table must therefore reflect both of the degrees of freedom.

Returning to our example

The calculations associated with our example data are outlined below:

The mean of the variances for the adjuvantly treated groups is:

$$[(73.45)^2 + (77.36)^2 + (99.23)^2 + (86.27)^2 + (97.78)^2 + (61.87)^2]/6$$
$$= 42057.4/6$$
$$= 7009$$

The grand mean, i.e. the mean of the means, is 1554.86. The variance of the means is:

$$[(1763.28 - 1554.86)^2 + (1719.81 - 1554.86)^2 + .. + (992.75 - 1554.86)^2]/(6 - 1)$$
$$= 449853.44/5$$
$$= 89970.69$$

This is the estimate we called $s_{\overline{X}}^2$ To obtain an estimate for σ^2, we need to multiply this variance (the one that looks like a 'standard error') by the size of each group. In this case the size of each group, N_j, is 7. Therefore, this estimate for σ^2 is 629 796. The ratio of the two is 89.95.

These variance estimates appear in the ANOVA table under the column mean square (see page 81): the term mean square is just another name for the variance. It is derived from the fact that the variance is a sum of squared differences divided by the number of degrees of freedom left over after the estimates have been paid for in information coins.

The ratio of the two variance estimates appears in the column marked *F*-statistic, and the probability of observing this ratio when the null hypothesis is true, the overall treatment effect *p*-value, is given to the right of the *F*-statistic. All of the information pertinent to the test is summarized in the lower right hand corner of the table.

Since the null hypothesis assumes that all the sample populations (treated and control) were drawn from a single underlying population, each estimates the true mean and variance. The statistical question 'What is the likelihood of observing a ratio 'this big' between the variance of the means and the mean of the variances, given that the null hypothesis is true?' is the link in the bridge which associates that ratio with a probability measure.

The test is run as follows. First calculate the numerator for the test, i.e. the estimate of σ^2 based on the variance of means, 629 796. Then calculate the denominator for the test, i.e. the estimate of σ^2 based on the mean of the variances, 7009. The number of degrees of freedom for the numerator is 5 (6 treatment groups minus one estimate of the grand mean), and for the denominator is 36 (42 individual measures minus 6 estimates for each group mean or variance). Calculation of the ratio of the two variance estimates gives a value of 89.95. That ratio can then be linked with a probability using the *F*-table with 5 and 36 degrees of freedom. The critical value drawn from the ANOVA table on page 81 is 2.48. This means that, with this study design, the likelihood of seeing variances that are about 2.5-fold different when the null hypothesis is true is only

5%. The p-value for observing two estimates of the same variance which are 90-fold different is less than 0.0001 or less than 1 chance in 10 000! Based on these results we are willing to reject the null hypothesis and say that at least one of these groups is different from all the others.

Pairwise comparisons

The final question to be answered is 'Which of these treatments was active when compared with controls?' You could just as easily have asked 'Which of these treatments is different from which?': in either case you would still test an adjuvant-treated control against a non-treated control to see whether the adjuvant induced proliferation in autoreactive T-cells. Assuming it did, you would then drop the non-treated group to determine the effects of the test compounds. The ANOVA would actually be performed the same way, but the scientific questions are different. The *pairwise comparisons* used to answer those different questions will therefore also be different.

The tools used to make pairwise comparisons within the context of an overall ANOVA are numerous and complex. They differ according to the design of your study and how conservative you feel about making an error during any one of the comparisons: we cannot, therefore, outline all the possible scenarios in this book. For the sake of completeness, however, we will highlight the two which could be employed to answer the questions raised at the beginning of this section.

The first test allows comparison of any treatment group (usually the control) with all the others. The mathematics does not care which is which. You should, therefore, take care during the design phase to identify your control adequately and be sure that this is the only one used for comparison. In the foreword to this book, we raised the point that you could compare your bioassay results to a standard reference compound, to an untreated control, or maybe both. This is a case which has two possible controls and two sets of comparisons.

Assuming that you are really only interested in one set of comparisons, the test we employed to derive the table of comparisons shown on page 82 was *Dunnett's test*. This test adjusts the pairwise comparisons in a fairly conservative manner for each of the five (k–1) comparisons. The details of the adjustment are based on the results generated in the overall ANOVA, and are beyond the scope of this book. Graphically, the comparison scheme looks like a 'starburst', with the control comparator in the middle, and the individual comparators radiating outward (see Figure 21).

The other point raised in the foreword was that you might want to know which of these drugs 'works', and which works better than the others. This test would clearly have more possible comparisons than those performed in the case outlined above. In the case study with six groups there were five independent Dunnett's comparisons out of 15 possible. Determination of which is different from which requires all 15 of those questions to be answered. Given our concern over controlling the experiment-wise error rate, shouldn't any test which compares 15 pairs be more conservative than one which compares only five? In other words, we would need the means to be farther apart in the noise space. This more conservative, but more general, test is called the *Student–Neuman–Keul's test*. Again the details of the test are beyond the scope of this book, but, it is important to determine which question is to be answered **before** the experiment is performed so that the right tools are used to analyze the results.

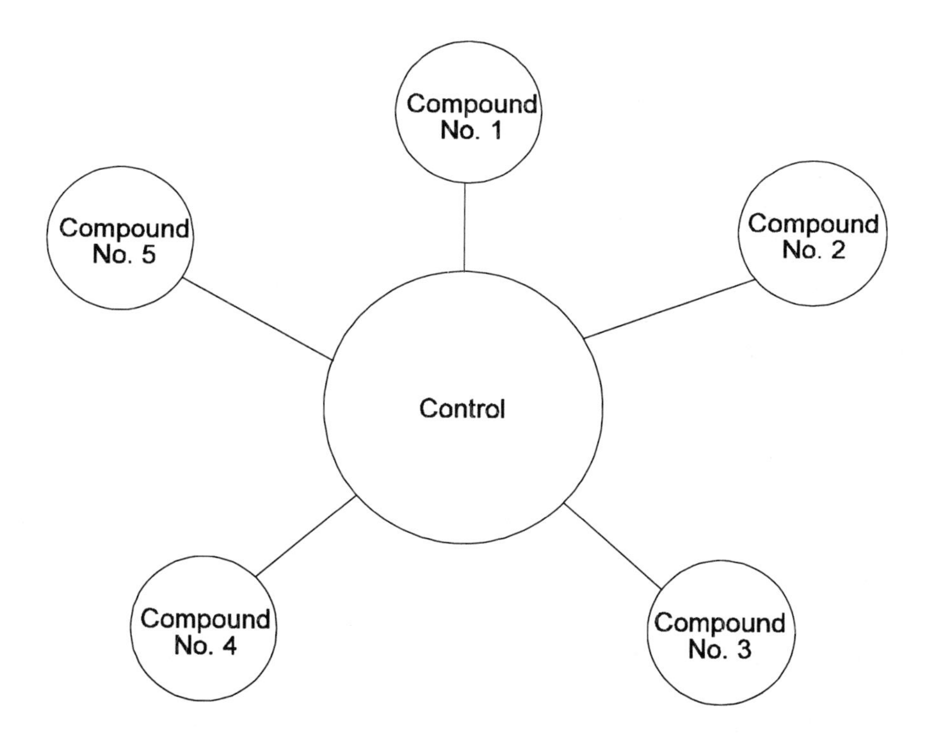

Figure 21

COMPARING TWO OR MORE PROPORTIONS: PROPORTIONS TESTS AND CHI-SQUARE (χ^2)

Terms you should learn:
- Two-sample proportions test
- Chi-square test
- Blinding
- Binomial distribution and Bernoulli trials
- Marginal totals
- Chi-square distribution
- Rule of five & Fisher's exact test

Concepts you should master:
- Blinding, bias and category assignment
- The null hypothesis, its assumptions and interpretation
- Proportions as random variables
- The pooled proportion as an estimator
- The variance of the distribution of proportions
- Degrees of freedom in the chi-square

Foreword

Up to now we have been developing tools with which to compare the average behavior of two or more populations, and our responses have been summarized as arithmetic means. But suppose we need to compare two or more populations **as a whole** with respect to their abilities to respond to a particular stimulus. This question is slightly different to those previously considered, which have involved measurement of a continuous response (e.g. blood pressure, cell counts, etc.) from each member of the sample population. In this case we establish, *a priori*, a level of response which we identify as 'positive' and ask 'What proportion of the population is positive, and is that proportion different from population to population?'

In essence, while each subject in your study contributes either 'yes' or 'no' to the overall answer, the sample population **as a whole** is the real experimental unit. From that sample proportion you will try to establish what the true proportion of responders would have been if you had tested, simultaneously, every subject in the underlying population. Using this estimate you can then ask the same inferential question as you asked about the means: 'How far away, in probability space, are the proportions observed in sample population 1 and sample population 2, and could they both be snapshots of the same underlying population?'

Case study

Pain in the neck has been shown to be a major debilitating condition afflicting the average mother of three. Based on extensive research into neck physiology, nerve conduction and child rearing, a model involving a pinched nerve in the neck has been developed which mimics this condition in female rats. When these animals are stroked along the back they writhe and squeal in pain. Although the number of writhes and squeals is difficult to quantify, it is possible to determine whether or not the animals are in pain, and they can therefore be assigned to one of two groups (happy or sad). It has been proposed that the proportion of rats in the happy group will increase when an active analgesic (trade name Pain-Away) is given to the rats 2 hours before stroking. Is this correct?

The experiment designed to test this hypothesis compares Pain-Away with three other analgesics to see (1) whether Pain-Away really works and (2) whether there is any evidence to suggest that Pain-Away is better than anything else. You determine that five groups (Control, Pain-Away, and three competitors) of 20 animals each is a large enough sample size to show any relevant differences in analgesia. Each animal in each group has been surgically prepared with a neck pinch, and each is randomly assigned to one of the treatment groups. The treatments are performed as follows: a rat is selected at random from each group (the experimenter does not know which rat is from which group, a technique known as *blinding*). He strokes it across the back, and declares it happy or sad. The process continues, 5 rats at a time, until all rats are tested.

The data

The data for your experiment are given in Table 19. Removing the control group allows the second question to be answered. The new data display is shown in Table 20; this yields an analysis called *a chi-square (χ^2) test*. It is analogous to the ANOVA in that it answers the same type of question about proportions that the ANOVA answers about means, i.e. 'In these multiple samples, is there enough evidence to suggest that at least one of them is not drawn from the same underlying population?'

Data analysis

The data are analyzed as follows. First, the control group is compared just with the group treated with Pain-Away (proportions 6/20 and15/20) to determine the likelihood that a single underlying population of pinch-

Table 19

Group	Happy	Sad	% Responding
Control	6	14	30
Pain-Away	15	5	75
Comparison 1	12	8	60
Comparison 2	10	10	50
Comparison 3	13	7	65

Table 20

Group	Happy	Sad	Total
Pain-Away	15	5	20
Comparison 1	12	8	20
Comparison 2	10	10	20
Comparison 3	13	7	20
Total	50	30	80

necked rats would have yielded two sample proportions 'this far apart'. The statistics are similar to those in the two sample *t*-test (Table 21). In this table there is an estimator not seen before, called the pooled estimate of the proportion. We will discuss it more fully below but briefly, it represents the best guess we can make about the true underlying proportion of happy rats when the null hypothesis is true. Another difference between this analysis and the two-sample *t*-test is that this test statistic is distributed like a *z*-statistic and not like a *t*-statistic. This is something to keep in mind.

Table 21

	Control		Pain-Away
Proportion responding	6/20 (30%)		15/20 (75%)
Pooled estimator of proportion		0.525	
Pooled SE		0.158	
z-statistic		2.848	
p-value		< 0.005	

The second analysis (Table 22) is a bit more difficult. This table is called a χ^2 contingency table.

Table 22

Group	*Happy* Observed/ Expected	*Sad* Observed/ Expected	Total
Pain-Away	15 / 12.5	5 / 7.5	20
Comparison 1	12 / 12.5	8 / 7.5	20
Comparison 2	10 / 12.5	10 / 7.5	20
Comparison 3	13 / 12.5	7 / 7.5	20
Total	50	30	80

$\chi^2 = 2.773$ df = 3; Critical value = 7.815; $p = 0.428$

This analysis yields a *p*-value similar to that derived from the ANOVA. It indicates the likelihood of obtaining a group of proportions this disparate, given the null hypothesis is true, i.e. knowing they were all drawn from the same underlying distribution. In the headings of the second and third columns, 'observed' is the result obtained, while 'expected' is the value expected if the null hypothesis were true. What this analysis quantifies is how far away from expectation the actual results really are. Expectation is, of course, defined by the null hypothesis.

The question and your experiment

The study began with a question about the efficacy of a standard drug (Pain-Away) and was expanded to a comparison of Pain-Away with other analgesics. A reasonable model of pain which mimics neck pain in the average mother of three was obtained and, using standard technologies, you decided that while the number of squeals and writhes in any single rat was not a good measure of pain, the proportion of happy rats in each test sample was a reasonable measure of efficacy. What you are asking is 'Does my model of analgesia really work, i.e. does pain actually decrease when I treat my rats with the wonder drug, Pain-Away?' If so, is there any difference between Pain-Away and the other three analgesics tested? What your experiment is asking is, '**proportionately**, are there more rats assigned to the happy category in the treated group than in the group of animals receiving a placebo?', and if so, '**proportionately**, do any of the other treatment arms yield significantly lower assignment levels than Pain-Away?'

Statistically, you are trying to determine the chances of observing a distribution of sample proportions as disparate as that obtained, given that these samples were all drawn from the same underlying population.

Assumptions

The scientific assumptions underlying the use of all proportions tests, including the χ^2, are:

(1) the true proportion of responders in the underlying population is not known;
(2) the number of rats observed in the happy group is a scientifically valid measure of efficacy;
(3) the underlying populations are truly homogeneous, i.e. only pain responses due to pinched necks are being measured in the assay.

The statistical assumptions required for this test are:

(1) the measures are dichotomous (e.g. all-or-none effects);
(2) the samples are drawn from the same underlying distribution;
(3) the sample populations are chosen randomly from the underlying population;
(4) the measurement and assignment tool is not biased or confounded with other effects (which is why the study was blinded);
(5) there are enough subjects per treatment arm to show differences in proportions which are deemed to be biologically relevant.

The test – the equation, what it means and the distance it measures

In general, the null hypothesis for the first question is given by:

$$H_0: \pi_1 = \pi_2 = \pi$$

The notation has changed to indicate that we are looking at proportions and not means: the Greek letter 'pi' is used to indicate the 'true proportions' of underlying populations, and the Roman alphabet (letter p) is used to indicate sample proportions, i.e. those actually observed. What this equation is actually saying is that the proportion of responders in the placebo control group is equal to the proportion of responders in the group treated with Pain-Away, and that both are equal to the underlying proportion of responders in all rats, π.

The equation which tests this assumption is another critical ratio:

$$z = \frac{p_t - p_c}{\sqrt{p^*(1 - p^*)(\frac{1}{N_t} + \frac{1}{N_c})}}$$

where

$$p^* = \frac{N_t \ p_t \ + \ N_c \ p_c}{N_t \ + \ N_c}$$

N_t and N_c represents the sample sizes of the two populations respectively (the subscript t indicates those treated with Pain-Away, while c indicates control). p_t and p_c represent the proportions of responders in the two sample populations respectively, and p^* is the weighted mean of the two, and is a *pooled estimate of the true proportion*. This is, in fact, simply the total number of successes divided by the total number of trials, and should be a reasonable estimate for π when the null hypothesis is true.

The first equation is just like the critical ratio calculated before for the two-sample t-test. It says that we are interested in a difference of two proportions ($\Delta = p_t - p_c$) which is normalized in some way for the noisiness of the estimates. If we consider $p_t - p_c$ as a single variable, Δ, this test (just like the two sample t-test) looks at a single realization drawn from the distribution of the random variable, 'differences in proportions'. Under the null hypothesis, i.e. assuming $\pi_1 = \pi_2 = \pi$, this distribution should have zero as its mean (just as in the t-test). In order to determine the distance from zero to the sample difference, and to see whether it is far enough away to allow the validity of the null hypothesis to be questioned, we need to know the expected spread around zero for the values of Δ when the null hypothesis is true. That noise measure is based on the *binomial distribution*.

The binomial distribution

A binomial event can have only two possible outcomes: positive and negative, happy and sad, heads and tails, etc. The expected value of the proportion of successes in N independent trials will be, on average, $N\pi$ where π is the underlying probability of success. It is important that each of these trials (in our case, each pinch-necked rat) is independent, and that the likelihood of success in each trial (rat) is equal. Each on–off, dichotomous test like this is called a *Bernoulli trial*, and the number of successes in N Bernoulli trials is a random variable. Since it is a random variable it must be associated with a distribution, in this case the binomial distribution, the variance of which is given by $N\pi(1 - \pi)$. The proportion of successes, as opposed to the number of successes, is also a random variable. It therefore has a distribution which has a variance given by $\pi(1 - \pi)/N$. The standard error of the proportion of successes, then, is just:

$$\sqrt{\frac{\pi(1-\pi)}{N}}$$

The proportions version of the critical ratio can be treated the same way as when we generalized the one-sample *t*-test to the two-sample *t*-test. We need an estimate of the variance of the population of differences in proportions when the null hypothesis is true, which should be, for consistency's sake, a weighted average of the two sample variances. That is easier to see as:

$$\text{STDERR} = \sqrt{\frac{\pi_t(1-\pi_t)}{N_t} + \frac{\pi_c(1-\pi_c)}{N_t}}$$

$$= \sqrt{[\pi(1-\pi)][\frac{1}{N_t} + \frac{1}{N_c}]}$$

The leap to the second line in this equation is derived when we assume that the null hypothesis is true, i.e. $\pi_t = \pi_c = \pi$. Therefore, **given the null hypothesis is true,** we can use this estimate of the noise.

In our actual test, we do not know the value of π: this is where p^* comes in. p^* is actually a weighted average, and the weights are given by the two sample sizes (N_t and N_c). Substituting p^* for π in the equation for the standard error and putting all these mathematical building blocks together, we arrive at the critical ratio given above.

The distance calculated in noise units (the difference in proportions divided by the noise) can now be associated with a probability distribution to give the likelihood measure, i.e. the probability value, we are really seeking. In this case, the critical ratio is distributed just like the *z*-distribution outlined in the one-sample tests of means above. The critical point which traces out a 95% confidence level for a two-sided test is 1.96. This means that the likelihood of seeing two proportions more than 1.96 standard error units away from each other **in either direction** is less than 5% when the null hypothesis is true.

We will work through the calculations needed during this 'assay verification phase' (i.e. the part of the experiment that answers the question of whether Pain-Away reverses neck pain in rats with pinched nerves) in detail below. First, however, we want to present an overview of the chi-square analysis.

The null hypothesis for the second question is given by:

$$H_0: \pi_1 = \pi_2 = \pi_3 = \pi_4 = \pi$$

This hypothesis states that the proportion of responders in each group is equal to any other and that all are equal to the same underlying proportion of responders, π. In other words, all sample populations were really drawn from the same underlying population ('effectively treated rats') and the differences observed in success rates are due only to random sampling.

The motivation for the chi-square test is just a generalization of the argument presented for the two-sample proportions test. The logic is as follows. If the null hypothesis were true, then the best estimate for π would be the total number of responders divided by the total number of participants (just like p^*), and there should be no major discrepancies observed from row to row of the χ^2 table (what is sometimes referred to as *row effects* in a χ^2 analysis). The total number of happy rats, given at the bottom of the second column in the χ^2 contingency table, is one of seven possible totals derived from this table (total number tested for Pain-Away, total number tested for comparison drugs 1, 2 and 3, total happy, total sad and the grand total). As a group, these totals are called *the marginals of the table* (because they are on the edges). The total number of happy rats divided by the grand total is therefore the best estimate of π when the null hypothesis is true. There should be $N_t\pi$ happy rats in the Pain-Away group. The same kind of result would be expected for the other totals. These calculations yield the expected numbers of successes in each group (the expected values in the χ^2 table above). If we were to sum all the differences between the expected and observed values in the table, we should get an overall discrepancy measure. If expectation is based upon the assumption of the null hypothesis, we would now need a means of associating our calculated discrepancy with a probability of observing it. This is the same paradigm used throughout all inferential testing. So what we want to do is normalize our discrepancy by a measure of its noisiness.

The equation for calculating the value of χ^2 is:

$$\chi^2 = \sum \frac{(\text{obs} - \text{exp})^2}{\text{exp}}$$

where the sum is taken over each of the cells in the χ^2 table.

The numerator of the sum is a squared discrepancy. The reasoning behind this is the same as that behind the use of the sum of squared deviations in calculations of variance – in order to obtain a positive sum, each of the components of that sum must be made positive. Although the absolute values could have been used, the arguments made in the case of the variance also hold for the χ^2 test.

The denominator in each component of the sum is the expected value derived for the cell in question. For the successes, it is the total number of successes divided by the grand total multiplied by the number of subjects tested in the group, while for the number of failures it is just its complement (Table 23).

Table 23

Group	*Happy* Observed/ expected	*Sad* Observed/ expected	Total
Pain-Away	15 / 12.5	5 / 7.5	20
Comparison 1	12 / 12.5	8 / 7.5	20
Comparison 2	10 / 12.5	10 / 7.5	20
Comparison 3	13 / 12.5	7 / 7.5	20
Total	50	30	80

In this case 50/80 = 62.5% or 0.625, which is the best estimate of π. The best estimate of the number of happy rats expected following treatment with Pain-Away, if the null hypothesis is true, is 20 times 0.625, or 12.5. However, the χ^2 sum, composed of all the discrepancies found in your table, is a random variable just like all the other test statistics discussed on the preceding pages. It must, then, have a distribution from which a probability measure can be derived. Fortuitiously, it does and it is tabulated in most statistics references and software packages.

There is still one more point to be addressed. Given a fixed marginal (i.e. knowing that 20 animals were tested), specification of the number of successes automatically gives the number of failures in that particular sample. The fact that the table has fixed marginals means that the cells are not independent, and again raises the question of degrees of freedom. In this case, the degrees of freedom are a function of the number of independent cells in the table, and not of the number of subjects tested. A table with R rows and C columns has (R – 1) × (C – 1) independent free spaces to be filled before all the rest are automatically calculated. The χ^2 distribution is therefore described by (R – 1) × (C – 1) degrees of freedom. It should therefore be obvious that the larger the χ^2 table, the more degrees of freedom it has, and the more entries there are in the sum. Thus, as the table grows, the sum will also grow, and the critical value on the total χ^2 needed to reject the null hypothesis also increases. In other words, the more degrees of freedom you have the greater the total sum must be to reject the null hypothesis. This is opposite to the case for the *t*-distributions, where the fewer the degrees of freedom the larger the critical ratio had to be to yield the same the *p*-values. The reason is that in the case of the *t*-distribution, a lower number of observations increases the uncertainty of the estimate of the true variance, and this uncertainty was accounted for by making the critical points larger.

Returning to our example

The calculations associated with the first test are shown in Table 24. The null hypothesis assumes that the two sample populations (treated and control) were drawn from a single underlying population, and hence, will have the same proportion of positive responses. The statistical question, then, is 'What is the likelihood of observing a difference in proportions this big (= 0.75 – 0.30) given the null hypothesis is true?' The test will associate that difference with a probability measure.

The test is run as follows. The numerator for the test, i.e. the difference, is 0.45. The pooled estimate of the proportion, p^*, is estimated by adding up all the responders and dividing by the total number of independent trials (rats); this gives a figure of 0.525. The pooled estimate of the standard error, derived from p^*, is 0.158. The distance metric requires calculation of the ratio of the difference and the standard error, which has a value of 2.848. Associating that with a likelihood of occurrence by using the *z*-distribution yields a probability of seeing a difference this big, given that the null hypothesis is true, as less than 0.005, or 5 chances in 1000. It is therefore probably fairly safe to reject the null hypothesis and to conclude that Pain-Away actually does alleviate severe neck pain.

The calculations associated with the second test are outlined in Table 25.

Table 24

	Theory	Disease model
Proportion responding	p_c & p_t	6/20, 15/20 (30%, 75%)
Pooled estimator of proportion	$p^* = \frac{N_t p_t + N_c p_c}{N_t + N_c}$	21/40 (52.5%)
Pooled SE	$SE = \sqrt{[p^*(1-p^*)][\frac{1}{N_t} + \frac{1}{N_c}]}$	0.158
z-statistic	$z = \frac{p_t - p_c}{\sqrt{p^*(1-p^*)(\frac{1}{N_t} + \frac{1}{N_c})}}$	2.848
p-value		<0.005

Table 25

	Theory	Disease model
Proportion responding	p_1, p_2, p_3, p_4	15/20, 12/20, 10/20, 13/20 (75%, 60%, 50%, 65%)
Pooled estimator of π	total responding/grand total	50/80 (62.5%)
Chi-square sum	$\chi^2 = \frac{(\text{obs} - \text{exp})^2}{\text{exp}}$	$(15-12.5)^2/12.5 + (5-7.5)^2/7.5 + \ldots (7-7.5)^2/7.5 = 2.773$
df	$(R-1) \times (C-1)$	$3 \times 1 = 3$
χ^2-statistic critical value		7.815
p-value		$p = 0.428$

The null hypothesis assumes that all the sample populations were drawn from a single underlying population, and hence, that they all have the same underlying proportion of positive responses. The statistical question is, therefore, 'What is the likelihood of observing a pattern of sample proportions this disparate (75%, 60%, 50%, 65%), given the null hypothesis is true?'

The test is run as follows. The best estimate of π, based on a weighted average of happy rats, is 50/80 = 62.5%. Given this estimate of π, should four proportions drawn at random from the distribution of all sample proportions (of samples of size 20) have this pattern of response? The distance metric requires calculation of the sum of normalized discrepancies (i.e. the ratios of the squared differences between observed and expected values normalized by the expected values), which is 2.773 noise units. That sum is associated with a likelihood by using a χ^2 table with three degrees of freedom. The critical value which ensures at most a 5% error rate is 7.815. In other words, the sums of normalized ratios greater than 7.815 will occur, by chance alone, less than 5% of the time, when the null hypothesis is true. In the case described above, a sum of discrepancies greater than 2.773 occurs about 43% of the time, and this can occur by chance alone. Most people would be unwilling to reject the null hypothesis knowing that they would be wrong 43% of time, and these results therefore do not suggest that Pain-Away is better, or worse, than any other analgesic.

Caveats and extensions

There are two more issues of which you need to be aware when using the χ^2 test. First, none of the theory that we have covered expressed a requirement for the use of dichotomous responses in the columns of the χ^2 table. All that was involved was the addition of a set of normalized discrepancies, and checking to see whether their total was 'far enough' away from expectation to suggest that the null hypothesis should be questioned. This suggests that the same logic should apply to any number of well-defined categories for which expected values could be derived, and this is in fact the case. Consider the example shown in Table 26. For the first type (category) of response, observations were made in five subjects from a sample of 20 (25%) from group A, in seven of 35 (20%) from group B, in 10 of 25 (40%) from group C, and in six of 15 (40%) from group D. Could these sample proportions have come from a single underlying population in which 29.5% (28 of 95) were type 1 responders? The same question could be asked with respect to types 2, 3 and 4. The sum of all the discrepancies derived from this table does not depend upon the simple proportions observed with dichotomous responses. Rather, it only needs a complete set of proportions for all categories to enable the calculation of expected values to be undertaken. Once all the machinery is in place, the logic of adding up all deviations from expectation remains the same. In this case, there are four groups and four categories of response, i.e. it is a 4×4 χ^2 table, and there are $3 \times 3 = 9$ degrees of freedom. Once the sums have been calculated, the χ^2 distribution allows us to calculate a probability of seeing a total normalized discrepancy 'this big' when the null hypothesis is true (i.e. when there are no row effects, and all the groups are just snapshots of the same underlying population).

Table 26

Group	Type 1 Observed/ expected	Type 2 Observed/ expected	Type 3 Observed/ expected	Type 4 Observed/ expected	Total
A	5 / 5.9	7 /5.5	etc.	etc.	20
B	7 / 10.3	9 /9.6	etc.	etc.	35
C	10 / 7.4	5 /6.8	etc.	etc.	25
D	6 / 4.4	5 /4.1	etc.	etc.	15
Total	28	26	20	21	95

The second point is that there are limits to the χ^2 technology. For theoretical reasons that are beyond the scope of this book, the χ^2 table cannot be too sparse. By that we mean that the table cannot contain too many cells that are either empty or contain a small number of counts.

Suppose a set of categories in a particular assay yields only small proportions of responders. Since we are considering only whole number events (i.e. the number of responders in a given category), one responder who deviates from expectation in either direction, by chance alone, will alter your final total by adding to the discrepancies observed. However, the χ^2 calculation requires the use of normalized discrepancies, and the normalization factors used are the expected values themselves. Small sample cells would therefore be inordinately weighted by these smaller expectations to be fair. If a data set does have too many of these small sample cells the problem is probably not insurmountable, and the χ^2 analysis will probably work. Too many of these cells, however, could cause trouble. The statistician has a rule of thumb, called the *Rule of Five,* which he/she applies to the χ^2 table before proceeding with the analysis.

This rule says that if 20% of the cells in the table contain expected counts less than five, use of the χ^2 test should be reconsidered. An analysis tool known as *Fisher's exact test* should be used instead. This numerical test, which can be found in most statistics software packages, analyzes the likelihood of the pattern of cells observed (or something even more extreme), given fixed marginals. This probability is calculated exactly by enumerating all the possibilities for the given table structure. Therefore, even when the χ^2 tool is inappropriate you can still analyze your results.

DISTRIBUTION-FREE MEASURES: NON-PARAMETRIC STATISTICS

Terms you should learn:

The sign test
The Wilcoxon signed rank test
The rank sum test
The Kruskall–Wallis test

Concepts you should master:

Parameters and distributions
Sign test and the binomial distribution
Signed rank and the paired *t*-test
Rank sum and unpaired *t*-test
Kruskall–Wallis and the one-way ANOVA
The null hypotheses, their assumptions and interpretations

Foreword

So far we have discussed statistical terms such as means, standard deviations and variances as population measures, as if we really know what we are talking about. Each of the case studies presented above assumed the existence of an underlying statistical population, the distribution of which looked like a normal (i.e. bell-shaped) curve, and assumed that the sample means and standard deviations, the measures of location and spread derived from random snapshots, were somehow representative of that distribution. Suppose the underlying statistical population yielded sample data that did not fit this image of a perfect distribution. In that case a mean calculated from the sample data need not necessarily represent the underlying statistical population and may not, in fact, be the most appropriate value to use when comparing target populations. A perfect example of data with a non-normal (i.e. non-bell shaped) distribution is scores on fixed intervals. This situation is described more completely when we recognize that the mean and variance of a normal population are just representatives, or *parameters,* of that distribution. We have assumed that the population can be described exactly using just these two values. That incredible feat was accomplished using complex mathematical formulae based upon these parameters which trace out theoretical curves. The class of inferential statistics that takes advantage of these complex formulae is called *parametric statistics.*

When assumptions about the shape of the underlying distribution fail, or if there are serious doubts, given the randomness of the sampling tools, about the shapes and sizes of these distributions, it would be nice to have a set of tools with which populations can be compared and described. Such tools are called *distribution-free* (i.e. assumption-free), or *non-parametric statistics*, because they do not require μ and σ the true mean and standard deviation to be used, estimated or tested.

In this section we will outline four non-parametric analogs for the paired *t*-test, the unpaired *t*-test and one-way ANOVA, parametric methods which were described above. There is no non-parametric analog for the one-sample *z*-test since that test depends intimately on the shape of the standard normal Gaussian curve. Without the standard normal curve there really is no *z*-statistic at all. In fact, the reason we developed the one-sample *t*-test in the first place was because we had to estimate σ from the sample data. There are also some distribution-free methods analogous to the tools used in correlation and regression (see Chapter 3). For the sake of consistency they will be presented under that topic.

Case study (the sign test and the Wilcoxon signed rank – for paired samples)

It has been suggested direct sunlight weakens the protein structure of the hair, and that repeated exposure destroys the sheen and luster so prized among the 'beautiful people' on the resort beaches of the world. A study is designed similar to that described above for the study of health promotion to reduce the risk of cardiovascular disease, in which green-eyed rats are sent to the Caribbean for exposure to sun and coconut baths, in an attempt to determine whether multiple exposures to the Caribbean sun result in the loss of tensile strength in the hair. The results obtained are shown in Table 27.

There are two things to notice about this table. First the data are ranked – the details of how this is done are presented below. Second, the ranks are also 'signed', i.e. they are given a direction. The important thing to notice here is that rank for the zero value (rat 3) is discarded, and remaining ranks are then determined by first direction and then magnitude. Those two aspects of the procedure establish the difference between the sign test and the signed rank test.

There are two positive differences, and the sum of the positive ranks is 5. There are seven negative differences, and the sum of the negative ranks is 40. The sample size is actually 9, since rat 3 was discarded. The

p-value for the sign test is 0.095, while the p-value for the signed rank test is < 0.05. How were these calculated and why are these two values so different?

Table 27

Rat	Tensile strength of hair: After (g)	Tensile strength of hair: Before (g)	Difference (g)	Signed rank
1	3.0	6.7	–3.7	–5
2	6.0	7.1	1.1	1
3	5.8	5.8	0	–
4	1.5	7.1	–5.6	–7
5	6.7	4.0	2.7	4
6	5.3	11.8	–6.5	–8
7	6.2	8.2	–2.0	–2.5
8	5.7	7.7	–2.0	–2.5
9	4.9	12.3	–7.4	–9
10	4.1	8.6	–4.5	–6

The question and your experiment

The study began with a question about the effects of sunlight on the protein structure of the hair. A model of measuring hair strength in green-eyed rats was developed that entailed pulling out the hairs and measuring the tensile strength of each strand. You measured these strengths before and after the animals were kept for 2 weeks in direct Caribbean sunlight. What you are asking is, 'In this model, does direct sunlight exposure result in protein degradation and the eventual loss of luster and sheen in the hair?' What the experiment is asking is, 'on the whole, is there a significant loss of tensile strength in the hair of green-eyed rats exposed to the Caribbean sunshine?'

Statistically, you are trying to determine the chances of observing a distribution of positive and negative differences (in both magnitude and direction) as disparate as that obtained, given that these samples were all drawn from a single underlying population centered around zero.

Assumptions

The statistical assumptions required to run the paired t-test are:

(1) the measures are continuous (e.g. not all-or-none effects, scores, etc.);

(2) they are drawn from an underlying distribution which is normal with a mean of zero;
(3) the sample population is chosen randomly from the underlying population;
(4) there is true pairing of the data.

The only difference between these assumptions and those which underlie the Wilcoxon rank sum test is that there is no need for normality in the underlying population (assumption 2). This method is distribution-free, and we must therefore alter the way we look at the underlying scientific assumptions.

Recall that the scientific assumptions underlying the paired *t*-test require the estimation of the true mean and standard deviation of the differences between pairs to enable a distance in a statistical space to be calculated and assigned a probability. The probability is based upon a distribution (the *t*-distribution). The basis of those calculations is the assumption that the underlying population is truly homogeneous, and that the true difference should be about zero.

The only part of these assumptions which pertains in this example is that the population is homogeneous and the distribution of differences is **centered** around zero. In other words, since these methods make no assumptions about the shape of the distribution of the statistical population, the standard parameters, mean and standard deviation, need not be estimated from the sample. Instead, it is assumed that the distribution of responses has a median of zero, i.e. the difference is as likely to go up as down. If that is true, then a pattern of responses, namely the positive and negative differences, should emerge. The distribution of the signs should be about even, and the sums of the positive and negative ranks should be about equal. This is the assumption tested in these particular non-parametric tests.

The tests – the same null hypothesis and its interpretation

In general, the null hypothesis is given by:

$$H_0: \ MEDIAN = 0$$

This says that without estimating the underlying mean and by assuming no effect of treatment or intervention, equal numbers of differences between pre-intervention values (tensile strength before) and post-intervention values (tensile strength after) should go up as go down. For these tests, there is no real distance measure analogous to that given in the paired t-test. Rather, there is an assumption about the underlying distribution of differences based upon a conservative assumption about the underlying science, i.e. no effect.

If the intervention (two weeks in the Caribbean) causes no real change in tensile strength, then any changes observed (i.e. any differences in the before and after measurements) must be due to random variation alone. If this is true, equal numbers of changes should go up and down, and the magnitude of these changes should be about the same. The distribution of pluses and minuses should therefore look like the distribution of heads and tails following multiple tosses of a fair coin, and the ranks of all the differences, independent of their sign (up or down, we only care about their magnitude) should show as many 'big' ups as 'big' downs. If we were to sum the two sets of ranks separately, the sum of the positives should be near the sum of the negatives.

The sign test

Consider only the direction of the changes. If the sample contains N subjects, then under the null hypothesis, assuming that sunlight has no effect on hair strength, you would expect about half your differences to go up and half to go down, i.e. the mean number of positives should be $N/2$. The variance of the distribution of the number of ups is derived from another statistical population known as the *binomial distribution*. This calculation is given as $Np(1-p)$, where p is the probability of going up. Since we assume no effect, and the chance of going up is the same as the chance of going down, $p = 0.5$. If the assumption about the lack of effect is correct, the difference between the results observed and those expected should only be attributed to noise, and the difference

$$Z = \frac{N \times 0.5 - \text{Actual positives}}{\sqrt{N \times 0.5 \times 0.5}}$$

is distributed as a standard normal z-statistic. A distance can therefore be assigned after all, and this distance yields a probability, which is the p-value of the sign test.

The sign rank test

The previous procedure ignores the magnitude of the differences and only looks at the distribution of their signs. The addition of extra information in the form of data about magnitude should produce a more 'powerful' test. (The idea of power is discussed in the section on experimental design later, but for now assume that we mean that one test is more likely than another to see the same size differences.) The sum of all the unsigned ranks is given by:

$$S = \frac{N(N + 1)}{2}$$

Under the null hypothesis, the sum of the positive ranks, S_P, should be about equal to the sum of the negative ranks, S_N. The signed rank test determines the number of possible pairs of sums, S_P and S_N whose total is S, and how likely it is that a pair as disparate as that obtained will occur by chance.

Returning to our example (Table 26)

There were 10 animals in the initial sample, and the differences between values obtained before and after the experiment range from 1.1 to –7.4. There is one difference of zero (rat 3). If the null hypothesis is true, and if we lived in a perfectly ordered universe, then this is exactly what we would expect. Therefore, rat 3 provides no information, and it can be dropped from our sample.

In the sign test, $Np = (9)(0.5) = 4.5$. There were two actual positives, and the square root of the variance $(9)(0.5)(0.5) = 2.25$ is just 1.5. The z-statistic is therefore $(4.5 - 2)/1.5$, or 1.67. The p-value associated with that statistic, derived from a standard normal curve just as in the z-test above, is 0.095.

For the signed rank test, there are only nine truly informative sample points. The sum of all the ranks, S, is 45. When we assign direction to the ranks, $S_P = 5$ and $S_N = 40$. What are the chances, given that the null hypothesis is true, of observing a difference between S_N and S_P 'this large'? The calculations have been done and the results are tabulated in most statistical books and software systems. In books they are arranged in (S_N, S_P) pairs, based upon the final sample size (in this case, 9). The result in the case described above is $p < 0.05$. What this says, is that the likelihood of seeing two sums at least as different as those observed (40 and 5) when the null hypothesis is true is less than 1 in 20.

A word of caution

Why is the *p*-value obtained using the signed rank test less than that obtained with the sign test? The reason is that the sign test ignores the **size** of the differences and looks only at the distribution of pluses and minuses. A lot of information is therefore discarded. If half of the differences were negative but all less than 1, and the remaining differences were all positive and all greater than 10, would you believe there was no effect of sunlight? Probably not, but the sign test would neither see these differences nor account for them. Assigning ranks adds extra information without adding any extra assumptions about the distribution.

A second point to consider is that we are dealing with the ranks of the differences. That means that if the largest difference was –1000 instead of –7.4 it would still be given rank 9. These techniques are therefore insensitive to possible outlying values. It is easier to include all of the data in an analysis and account for it with this kind of technique than to justify dropping a point just because it 'looks bad'. The decision to drop data is a very serious one, and should only be taken after expert advice from a statistician, and good scientific reason, such as possible contamination of data, are considered.

Case study (rank sum – for two independent samples)

Since sunlight induces degradation in hair quality, it is suggested that a protein-based shampoo should protect the prized sheen and glory.

Table 28

Hair tensile strength (g)	
Protein shampoo	Fake shampoo
3.0	6.7
6.5	4.1
5.9	5.8
6.5	7.1
6.3	7.0
8.3	5.4
6.2	8.2
8.7	4.7
9.9	12.2
6.5	7.1

In an experiment whose design is similar to that employed for the two sample *t*-test, two parallel groups of healthy Norwegian volunteers are taken to the Caribbean to investigate the effects of either a fake shampoo or a protein-based shampoo. The treatments are applied daily, after they have spent the day on the beach and just before they go to bed and the progress and compliance of the two groups is monitored closely. The results obtained are shown in Table 28.

The details of the calculations are presented below. But first look at the way the data are distributed (actual values and ranks):

3.0 **4.1 4.7 5.4 5.8** 5.9 6.2 6.3 6.5 6.5 6.5 **6.7 7.0 7.1 7.1 8.2** 8.3 8.7 9.9 **12.2**
1 2 3 4 5 6 7 8 10 10 10 12 13 14 15 16 17 18 19 20

Values for the fake shampoo are in bold type for the sake of clarity.

The median of the joint distribution of hair strengths is 6.5 (the 10th and 11th values in the array are 6.5 and 6.5, which yields an average of 6.5). This procedure asks the likelihood of seeing this sort of distribution for the fake shampoo (four below the median with ranks 2, 3, 4, 5 and six above with ranks 12, 13, 14, 15, 16, 20) when the null hypothesis is true.

The *p*-value for this test is > 0.05, i.e. the null hypothesis may not be rejected: there is no evidence for a real effect of protein shampoo on the tensile strength of hair in sun-exposed Norwegians. The likelihood of seeing a distribution of ranks at least this disparate is more than 5%. In fact it is much closer to 95%.

The question and your experiment

The study began with a question about the ability of protein to protect hair from the detrimental effects of sunlight. The results of the first experiment showed that sunlight was, in fact, quite damaging to hair. The second experiment was designed to try and determine whether protein-based shampoos can actually reverse that trend, or at least provide some protection against the effects of strong sun. A model was developed in which 20 normal Norwegian blond volunteers were exposed to the Caribbean sun for 2 weeks. Half of the subjects were assigned, at random, to a protein treatment and half were treated with fake shampoo, and the tensile strength of the hair from each of the volunteers was measured at the end of the 2 weeks. What you are asking is, 'Does protein-treated shampoo protect against damage induced by direct sunlight and return the sheen and lustre to the hair?' What the experiment is asking is, 'Is there any difference in the tensile strength of hair between Norwegian blonds treated with protein-based shampoos and those receiving a fake treatment (controls)?'

Statistically, you are trying to determine the chances of observing a distribution of strengths for the treated group within the joint distribution as disparate as that observed. The basis of the question is, of course, the assumption (the null hypothesis) that these samples were drawn from the same underlying population as those derived from the control group.

Assumptions

The statistical assumptions required to run the *t*-test are:

(1) the measures are continuous (e.g. not all-or-none effects, scores, etc.);
(2) they are drawn from underlying distributions which are normal;
(3) the sample populations are chosen randomly from the underlying populations.

The only difference between these assumptions and those which underlie the rank sum test is that there is no need for normality in the underlying population (assumption 2). Again, this method is distribution-free.

The scientific assumptions underlying the *t*-test are:

(1) the true means of the underlying populations involved will be estimated;
(2) the true standard deviation of those populations will be estimated, assuming equal variance, i.e. homoscedasticity;
(3) the difference of those means will be estimated assuming there is a single underlying population which is truly homogeneous.

The only assumption which pertains in this case is that the underlying population is homogeneous. This method makes no assumptions about the shape of the distributions of the statistical populations, so we do not need to estimate the means and pooled standard deviation. As was the case for the signed rank test, we are going to look at the distribution of responses and only assume that the tensile strength values are drawn from a joint distribution with the same median. If all that is true, then a pattern of responses should emerge, namely the sums of the ranks of the treated and control groups should be about equal. That is all we are testing in this particular non-parametric test.

The test – the null hypothesis and its interpretation

In general, the null hypothesis is given by:

$$H_0: \; MEDIAN_1 = MEDIAN_2 = MEDIAN$$

What this is saying is that without estimating any population parameters, and by assuming no effect of treatment or intervention, the distribution and ranks of tensile strengths will be similar for both treated and untreated hair. Again, there is no real distance measure analogous to that used for the unpaired *t*-test. Rather, we are assuming that the underlying joint distribution of ranks is based upon a conservative assumption of no treatment effect.

If the intervention causes no real change in our measurements, i.e. there is no effect, then any disparities in the ranks must be due only to random variation. One would suspect that equal numbers of the treated hair strengths should be 'large' as are 'small', and that the same holds true for the fake shampoo controls. Ranking all the values in a joint distribution should therefore show as many 'big' numbers from members of the treatment groups as from the controls. The same could be said for 'moderate' and 'small' values.

To perform this test, we again need the sum of ranks for each group. For a total of N subjects in a sample, the sum of all the ranks would be:

$$S = \frac{N(N + 1)}{2}$$

If all the subjects from one of your groups were ranked at the bottom of the joint distribution, their sum would be S. That is the smallest sum any group could accrue from the joint distribution, and the likelihood of getting all the subjects from one group as your lowest values, given the null hypothesis is true, is very, very small.

In fact, under the null hypothesis, the sum of the treated ranks, S_T, should be about equal to the sum of the untreated ranks, S_U, and the question this test asks is, 'how many possible pairs of sums, S_T and S_U are there, given sample sizes N_T and N_U, and how likely are we to observe a pair as disparate as that actually observed?

Returning to our example

There are 10 subjects in each of our samples. The smallest possible sum of ranks (1–10) is 55. Therefore, the most disparate pair of sums one could possibly observe would be 55 (1–10) and 155 (11–20). Summing the values for the treated and untreated groups separately gives:

$$S_T = 104 = (2+3+4+5+12+13+14.5+14.5+16+20)$$

and

$$S_U = 106 = (210-104).$$

What are the chances, **given that the null hypothesis is true,** that we would observe two values like those shown above for the sums of S_T and S_U? The calculations required have been done, and are tabulated in most statistical books or on-line analysis packages. They are arranged in N_T–N_U pairs. The calculations are based upon all the possible combinations of sums one could observe when there are N ($=N_T + N_U$) total subjects in the final joint sample size. The tables give you the critical pairs, i.e. the minimum disparities, which yield probabilities less than 0.05, 0.01, and 0.001. This is the p-value for your test .

In this case, $p > 0.05$. What this says is that the likelihood of seeing two sums like this (104 and 106), given the two sample were drawn from populations with the same median tensile strength, is fairly good (you cannot get much closer).

Case study (Kruskal–Wallis – for several independent samples)

Since the protein-based shampoo cannot protect hair adequately from the effects of exposure to intense sunlight, collagen-based shampoos are now going to be tested in 40 subjects, randomly assigned to four groups of 10, who will receive either a fake shampoo control, or one of three collagen-based formulas. This design should be familiar, since it is like that employed for the one-way ANOVA. Again, the treatments are applied every night after the subjects have spent the day on the beach.

The results obtained are shown in Table 29. We will forego the calculations needed to perform this test, but it is easy to imagine that they are simply an extension of the rank sum test. The generalizations are conceptually easy but computationally intimidating. We will, however, work through the conceptual interpretation of the null hypothesis, what it means, and how the results from a Kruskal–Wallis test are interpreted.

Table 29

Tensile strength of hair (g)			
Collagen 1	Collagen 2	Collagen 3	Fake Shampoo
8.7	5.3	7.0	5.7
9.6	4.9	9.5	4.6
5.8	4.8	9.9	6.8
7.1	5.3	8.5	5.1
7.8	5.8	8.3	6.8
6.3	4.4	7.3	6.4
6.8	6.1	6.2	6.2
7.7	4.7	9.7	5.7
6.2	6.0	10.9	5.2
6.9	5.0	8.5	6.1

The test – the null hypothesis and its interpretation

The null hypothesis for Kruskal–Wallis is related to that of the rank sum test in the same way as the null hypothesis for the one-way ANOVA is related to that of the unpaired *t*-test. For this example, the null hypothesis is given by:

$$H_0: MEDIAN_1 = MEDIAN_2 = MEDIAN_3 = MEDIAN_4 = MEDIAN$$

This says that without estimating any of the sample means, and by assuming no effect of treatment or intervention, the distribution of the ranks of the hair strengths from each of the groups will be similar, and could all be considered to be drawn from the same underlying population which has but one median. In other words, the ranked hair strengths will be randomly distributed about their own measure of central tendency, and no group will dominate either the high or low end of the scale. Thus, any disparities in the ranks must be strictly due to random variation.

Generalizing the equations for the sums of ranks, we could envision a world in which the four sums should all be about equal. This would certainly be the case if the null hypothesis holds. Instead of two sums, in this case we have four, and rather than pairs of sample sizes yielding our probability measure we have groups of four. The idea is the same, however: under the null hypothesis, the sum of the ranks for each of the groups (call them S_1, S_2, S_3, and S_4) should all be about equal, and the question this test asks is, 'how many possible sets of four sums are there, given the four sample sizes N_1, N_2, N_3, and N_4, and what is the likelihood of seeing sums as different as those observed?' In this case, each of the sample sizes is 10, so what you are specifically asking is, 'How likely is it that I will see these four rank sums within a joint distribution of size 40, given there are 10 subjects in each sample, and assuming that no treatment had any effect on hair strength?'

These calculations are quite complex and are usually performed on a computer. However, the conceptual framework of these calculations is easy to understand. Once the initial analysis is completed, the most you can infer is that at least one of your groups comes from a distribution different from the others. However, you can perform similar pairwise comparisons in this milieu in the same way as those performed in the analysis of data with a one-way ANOVA. The strategies used to control for the multiple comparisons problem, e.g. Dunnett's test, Student–Neuman–Keuls, etc. were derived to control for the experiment-wise error rate. Nowhere in their execution did they explicitly require the pairwise testing to be done with an unpaired *t*-test. In the non-parametric case we can therefore use Kruskal–Wallis as the analog for the one-way ANOVA and the rank sum test as an analog for the unpaired *t*-test, and derive an analysis strategy which is distribution-free but statistically sound.

3 ESTIMATION

Estimation resembles description except that you assume that an underlying relationship between two or more variables exists and that it can be specified in a mathematical formula. These relationships are termed *mathematical models*. Any relationship derived from your data is implicitly based upon the selection of the correct model. In fact, without even knowing it, you have been dabbling in modeling throughout the entire inferential statistics sections.

Other models are also constantly in use in research situations. For example, suppose two characteristics of a population are thought to be related proportionally. An example that has been used for time immemorial is that of height and weight. If you believe (for reasons we will explore below) that height and weight are directly proportional, i.e. for every extra inch in height a subject will weigh an average of an extra three pounds, you are assuming that height and weight are *linearly* related. Mathematically, you are assuming that the height–weight relationship (i.e. model) can be described by a formula that looks like: (see Figure 22). Weight should, however, be proportional to the volume

$$\text{Wt} = (\text{proportionality constant}) \bullet \text{Ht} + \text{baseline}$$

displaced by a body, and not to height. The units of volume are cubic inches while height is measured in linear inches. How can we get away

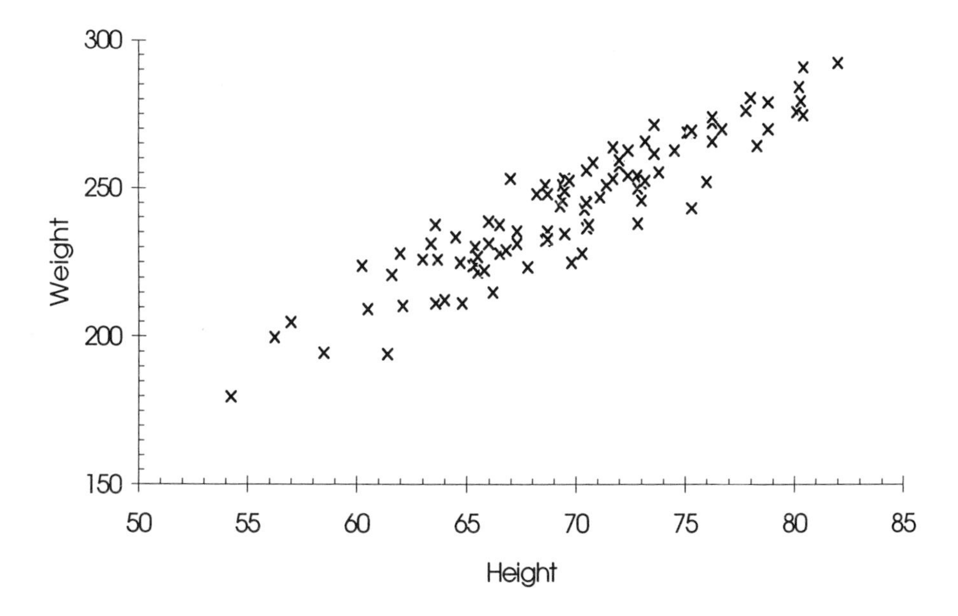

Figure 22 Height–weight relationship

with this (over-)simplification? The assumption underlying this model is that the human body is a cylinder, and that the volume of a cylinder is strictly proportional to its height ($\pi r^2 \cdot Ht$). The assumption implies that the radii of the human population are fairly constant, and that height is the determining factor in the calculation of weight. What you are trying to do statistically, therefore, is to determine from the observed data set, an estimate of the proportionality constant in that relationship, which represents the fixed radii of all humans and the density of human tissue.

The classic example of height and weight which we provided, though, is not, strictly speaking, a candidate for linear regression. In this chapter we will tell you why.

We will define the set of tools used to explore these kinds of relationships and help you understand what they can and cannot do. We will also look at the confidence that can be placed in these estimates, how these confidence regions can be determined and what they mean in the real world of experimental science. Finally, we will talk about how an experiment might be performed in order to optimize the information derived from it.

DATA RELATIONSHIPS: ASSOCIATION AND CORRELATION

Terms you should learn:

Scattergram
Bivariate normal
Correlation coefficient
Coefficient of determination

Concepts you should master:

Correlation as a measure of association
The correlation coefficient and what it means
Test of correlation coefficient
The null hypothesis and what it means
Association vs. cause and effect

Case study

The A22-Natasha amoeba causes a dreadful CNS disease, but these cells can be identified by their carriage of a unique and constant marker, the Boris-X1 receptor. It is widely believed that the concentration of Boris-X1 receptors on the surface of the A22-Natasha amoeba is fairly constant over the entire population of these creatures. This assertion is equivalent to saying that the number of receptors on a cell is proportional to the surface area of the cell, but an equally valid model of receptor expression is that the number of receptors per cell is independent of the surface area, i.e. the **number** of receptors, rather than the **concentration**, is constant and independent of cell surface area. If that were the case, then the genetic machinery which manufactures this particular receptor is totally independent of that which monitors the size of the cell. You have been asked to design an experiment to see whether the former hypothesis is correct. You therefore develop a new high frequency laser assay which can measure accurately the fluorescence of an object such as a cell once it has been tagged with a reactive dye. You then need to determine whether there is an increase in reactive fluorescence with target size. The following experiment is performed. A22-Natasha amoeba are grown in vitro for 2 days until they are in the exponential growth phase. One thousand cells are extracted at random and mixed with squirrel anti-Boris antibodies. Moose anti-squirrel antibodies tagged with reactive dye are then added to the mixture, and the volume and fluorescence of each cell is recorded.

The data

The data obtained in the experiment are given in Figure 23. This type of graph is called a *scattergram*. The *x*-axis shows cell volume while the *y*-axis is the intensity of cell fluorescence. The shape of the graph shows that the fluorescence intensity seems to fan out from left to right. A good statistician will recognize this as the graphical fingerprint of a heteroscedastic population: the variance increases with the magnitude of cell volume. Why should this be so?

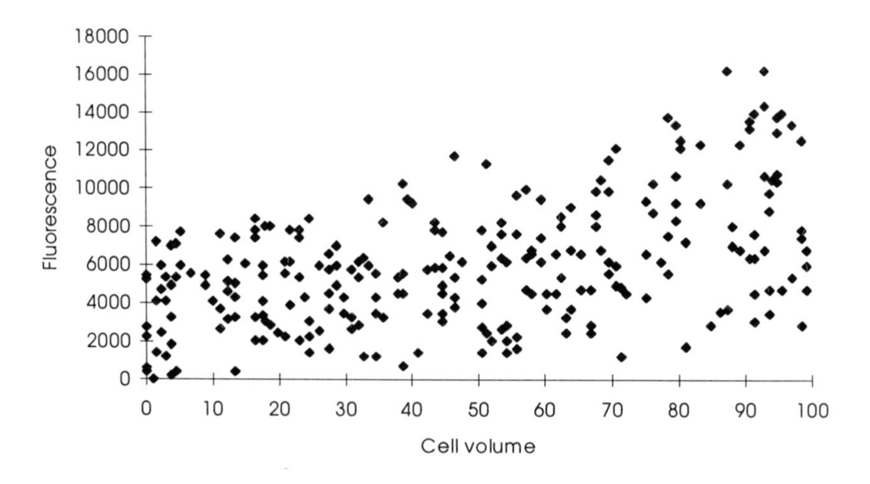

Figure 23 Scattergram of cell volume vs. cell fluorescence

Although you are trying to investigate whether the concentration of Boris receptors is constant on the surface of A22-Natasha amoebae, you are measuring cell volume, not surface area. Surface area, however, is proportional to the 2/3 power of the volume (μm^2 vs. μm^3). Therefore, you should not be too surprised that the curve of fluorescence vs. volume is not exactly proportional. It is possible to transform the data in a manner that is both statistically correct and biologically reasonable: after taking the 2/3 power of the *y*-axis data, a new plot can be created like that shown in Figure 24. Now that's better!

The question and your experiment

The study began with the suggestion that the concentration of Boris-X1 receptors on the surface of A22-Natasha amoebae is constant. An experiment was designed to measure the fluorescence of a specific dye

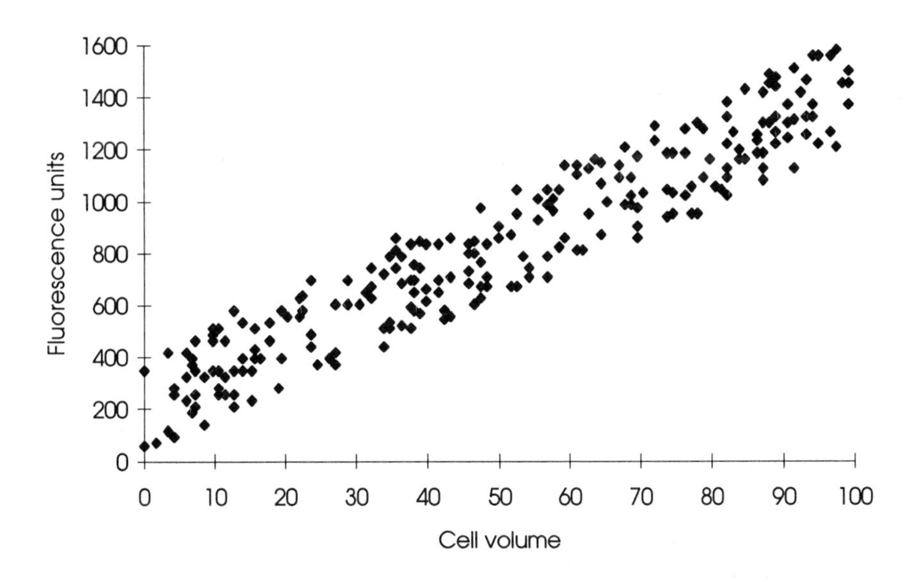

Figure 24 Data from Figure 23 transformed as described in text

marker on a randomly selected sample of target cells. The data obtained were plotted on a graph (Figure 23), but because these data showed severe signs of heteroscedasticity, they were transformed to produce the data set shown in Figure 24. What you are asking is, 'Do the A22-Natasha amoebae show a constant concentration of Boris-X1 receptors on their surfaces?' What your experiment is asking is, 'on the whole, i.e. **over the entire sample**, is there a proportional association between the cell surface area and the fluorescence observed in this random sample?' Statistically, you are trying to determine the chances of observing a proportional response as marked as the one observed, given that this sample was drawn from a single underlying population in which no proportionality exists, i.e. the case wherein receptor expression and cell size are unlinked (a fairly conservative assumption). The statistician calls this measure of association between these two random variables *correlation*.

Assumptions

To estimate and test the correlation between two random variables (usually, but not always, measured within a single population), one must assume that:

(1) the two measures are random variables, i.e. that neither is used to assign subjects to treatment groups or categories;
(2) each measure is continuous and not categorical or binary (i.e., is not 'all or none');
(3) each random variable is distributed normally and that together the random variables are drawn from an underlying *bivariate normal population*. By this we mean that the joint distribution of the two variables taken together results in a normal Gaussian distribution for each.

Calculation of the correlation coefficient and the coefficient of determination

The degree of association between two random variables (in this case fluorescence and cell surface area) is called the *correlation coefficient*, and is denoted by ρ (the Greek letter rho) for the population and r for the sample: r, which is called the Pearson correlation coefficient, is, in fact, a random variable, derived from other random variables, and is an estimate of the true value ρ. We will thus calculate r from a sample, estimate ρ, obtain a confidence interval about that estimate, and even formulate tests for ρ as we did for the mean, μ, for proportions, π, etc.

The equation for calculating the sample correlation coefficient can be expressed in two equivalent forms:

$$r = \frac{\Sigma(X_i - \overline{X})(Y_i - \overline{Y})}{\sqrt{\Sigma(X_i - \overline{X})^2(Y_i - \overline{Y})^2}}$$

or

$$r = \frac{\Sigma(X_i - \overline{X})(Y_i - \overline{Y})}{nS_X S_Y}$$

where n is the number of pairs measured and S_X and S_Y are the sample standard deviations of the two random variables taken individually. In our case n = 1000 A22-Natasha amoebae. But what do these two equivalent equations really say? The numerator is a *cross-product*. It says that when one of the values of X_i is less than its average, $\overline{X}$, i.e. when the difference is less than zero, then if the companion value of Y_i is also less than its average value, their product is positive. The same is true when both are greater than their respective averages. Therefore, if the random variables we call cell 'surface area' and 'fluorescence' range from their lowest to highest values in concert, a positive association between the two must exist. Similarly, if the value of X_i increases as the value of Y_i goes down, a negative association must exist.

The association must be normalized for the noise inherent in the underlying populations. This normalization is achieved by dividing the cross-product by a noise metric in the same way as in inferential statistics (see the discussion of *t*-tests outlined earlier). According to statistical theory, the denominators of the two equations, the first giving an explicit definition of the sum of products of squared deviations (something like a joint variance term), and the second showing explicitly that this is, in fact, the product of the two sample standard deviations, provide the noise metric. Once normalized, the correlation coefficient, r, can never achieve a magnitude greater than 1. Therefore, our measure of association is bounded by the values –1 and +1.

What do these normalized values mean? If x and y vary in perfect synchrony, meaning that every unit increase in x is accompanied by a proportional increase in y, then $r = 1$. Conversely, if for every unit increase in x, y decreases by a proportional amount, then while the two measures are still perfectly associated, they vary inversely and $r = -1$. The first phenomenon is called *a positive correlation* while the second is called *a negative correlation*. The degree to which these two measure deviate from that perfection due to random variation and noise is the degree to which the magnitude of r tends towards zero. When the correlation coefficient is zero (i.e. when the numerator is zero) you have perfect randomness and no association is therefore detectable between the two measures.

There is one last variable of association to be introduced before we proceed to the description of the tests that can be made using ρ. That statistic is called the *coefficient of determination*, and is just r^2. This coefficient is defined intuitively as the proportion of variance in the first random variable that can be accounted for by variance in the second. In other words, as one of your measures ranges from low to high, the degree to which the other ranges from low to high (or high to low for negative correlation), is measured by r^2.

Considering the extreme values of r; if r is either +1 or −1, r^2 is just 1. That means that changes in one variable can be accounted for perfectly by changes in the other. If r is zero, then so is r^2. That means that the second variable is completely independent of the first, and that irrespective of the values of x, the value of y is simply a realization from a normal distribution with its own mean and variance. In other words, the value of x is immaterial.

There is one last point which we wish to emphasize. This discussion has not used the terms independent and dependent variable. At no point has it been stated that one variable, y, depends upon the value of the other variable, x. That requires the derivation of a mathematical model that explicitly defines that relationship (like the one we derived for height and weight in our introduction). Rather, we are only saying that two random variables may be associated. Their assignment to the letters x and y is totally arbitrary, and the degree of association between the two, as measured by our ability to account for the variance of one by variance in the other, is the coefficient of determination. This is a rather subtle point, but is the foundation for the difference between correlation and regression, described in the next section.

The tests – the null hypothesis and its interpretation

The assumption that no association exists between two independent measures is a very conservative one indeed. You are positing that the underlying forces of biological nature which affect one measure exert absolutely no effect on the other. In statistical terms, that highly conservative assumption is embodied in the null hypothesis:

$$H_0: \rho = 0$$

In English, you are assuming that there is absolutely no association between the two random variables, and that the correlation coefficient of the underlying population is actually zero. We are going to return to the inferential paradigm and try to assess the likelihood of observing a sample correlation coefficient r whose magnitude is as large as the one actually observed, **given that no underlying correlation between the two measures exists at all**.

The equation which tests this assumption is:

$$t = \frac{r}{\sqrt{1 - r^2}} \sqrt{n-2}$$

where n is the number of pairs in the sample.

The test is another in the vast array of t-test tools available for estimating an underlying population parameter from sample data. Once these estimates have been obtained they can be tested to see whether they are significantly different from known fixed points. In a way this test is similar to that performed for the one-sample t-test described above. In this case we used a fixed point of zero, which corresponds to the highly conservative assumption that two measures are completely dissociated at a biological level.

Like any other t-test, however, we have to pay for information: the payment in this case is two degrees of freedom (information coins), one for each of the means estimated. Therefore, the degrees of freedom which will define the correct t-distribution for you is $n - 2$

Returning to our example

The summary statistics relating to the data obtained in the experiment described above are shown in Table 30.

Table 30

	Surface area	Fluorescence
Mean	49.55	8676.19
SD	29.38	4212.15

Although the actual calculations were performed using a standard computer software package, for the sake of clarity all the components needed for our determination of r are presented. The denominator for r, calculated as the sum of the cross-products of the raw data, is 123 752 967. The numerator, the product of 1000, 29.38, and 4212.15 is 102 343 703. Their ratio, the value of r, is 0.827. The p-value for this is < 0.01; r^2 is 0.684. These statistics tell you that there is a high degree of association between surface area and fluorescence ($r = 0.827$) and that the likelihood of seeing data this highly correlated, given that the null hypothesis is true, i.e., that there really is no association between these two measures, is less than 1 in 100. In fact, it is less than 1 in 1000. Furthermore, since $r^2 = 0.68$, 68% of the variation in fluorescence can be accounted for by the variation in surface area. However, this association is not a 'cause and effect' phenomenon, merely a measure of association. As silly as it may seem to the intuitive mind, we could also say that 68% of the variation in surface area can be accounted for by the variation in fluorescence. We address cause and effect as a phenomenon in the next section, in which we describe linear regression.

Extras: a non-parametric analog

Remember that we assumed that the data obtained had a bivariate normal distribution? Suppose that one or both of the random variables were not drawn from underlying populations with normal distributions. Then applying these techniques to estimate and test ρ would be sorely tried. However, just as parametric inferential tests can be augmented by using distribution-free methods of analysis, a non-parametric analog of Pearson's correlation coefficient is available to account for non-normality in your data. This non-parametric analog is called Spearman's rank correlation and is calculated as follows:

(1) Rank each of the (x,y) pairs from top to bottom. The ranking procedure is performed separately, first for x and then for y. Thus for 12 pairs of data, the largest value of x is assigned rank 1, the next rank 2, and so on, the smallest being rank 12. *Repeating* the procedure for the y variable yields a second set of ranks from 1 to 12. Therefore, when you have finished you should have pairs of ranks (X,Y) to replace your original data points (x,y). Ties are broken by taking the average of the ranks occupied by the tie.
(2) Subtract the two ranks Y and X and square the difference. If there is a perfect correlation, rank 1 for X will align itself with rank 1 for Y, and so on, so that all the differences will be 0.
(3) Add up all the squared differences and perform the following calculation:

$$R = 1 - \frac{6\sum(X-Y)^2}{n(n^2-1)}$$

When the sum of the ranks is zero the non-parametric rank correlation is 1. If X and Y are perfectly uncorrelated, $R = 0$. This estimate for ρ is valid when the initial assumptions about bivariate normality fail. The null hypothesis H_0: $\rho = 0$ can then be tested in the same way as above. The t-test is given by:

$$t = R\frac{n-2}{\sqrt{1-R^2}}$$

with n–2 degrees of freedom as before. Note that as R approaches either 1 or –1, t approaches infinity, meaning that the likelihood of seeing a sample rank correlation with magnitude near one, drawn at random from a population with a true rank correlation of zero is so small that you would feel quite comfortable in rejecting the null hypothesis.

If there are 'too many' ties this approximation will not work. What constitutes 'too many' depends upon your sample size and data, and you should review suspect data sets with an expert. If this should occur, then calculate the Spearman correlation coefficient using the same formulae you used to calculate Pearson's r.

DATA RELATIONSHIPS: MATHEMATICAL MODELS AND LINEAR REGRESSION

Terms you should learn:

Least squares approximation
Standard error of the estimate
Standard deviation of the regression
Goodness of fit
Multiple regression

Concepts you should master:

The line as a mathematical model
Linear regression and the one-way ANOVA
The slope and intercept as random variables
The null hypothesis and what it means regarding the slope
The null hypothesis and what it means regarding the intercept
Association vs. cause and effect
r^2 and accounting for the observed variance

Case study

You have been asked to explain why it is not possible to infer a 'cause and effect' relationship from the correlation results obtained above. We have already discussed why the measurement of two random variables cannot, by definition, yield variables that are classified as independent and dependent – a prerequisite for inferring cause and effect: if both cell size and receptor expression were controlled by a common, underlying genetic program, the experiment would have yielded these same results. Further studies are therefore necessary.

While looking for a control mechanism which can account for the varying levels of Boris-X1 receptors, you discover an apparently direct relationship between the concentration of a natural proteolytic enzyme, shermanase, and the expression of these receptors. The theory is that the receptors are masked by glycoprotein LHY-peabody-Q upon which shermanase acts. To determine whether this relationship exists, and whether the purported mechanism of receptor masking is real, the following experiment is performed. A22-Natasha amoeba are grown in vitro to exponential growth phase. Five sets of 50 amoebae are then extracted at random, and mixed with varying concentrations of pure shermanase. The amoebae are allowed to rest for 1 hour and are then exposed to squirrel anti-Boris antibodies followed by moose anti-squirrel antibodies tagged with the same reactive dye used in the previous experiments. The fluorescence of each cell is then recorded.

The data

The data for your experiment are given below in graphical form (Figure 25). The x-axis shows the concentration of shermanase, while the y-axis represents the intensity of cell fluorescence. These data give an estimated slope of 12.1 with a standard error of 0.16, and the estimated intercept is 123.3 with a standard error of 1.32. The p-values for testing the null hypotheses that these estimates are drawn from populations with means of zero are both < 0.01. The r^2 for the regression is 0.96. We will show you how to calculate these values and what they mean below. But first, we need to review and understand the questions asked and the experiment performed.

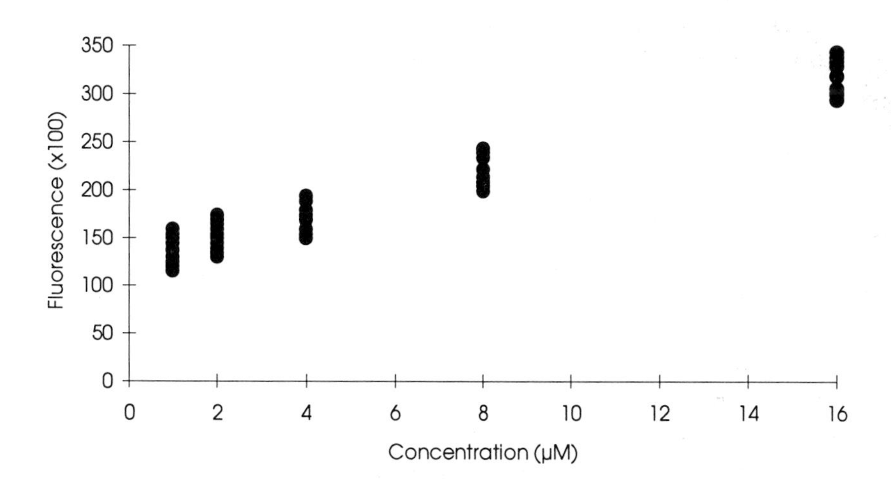

Figure 25 Plot of cell fluorescence vs. enzyme concentration

The question and the experiment

The study began with the question of an association between the size of A22-Natasha amoebae and the expression of Boris-X1 receptors. When a strong correlation was found, it became necessary to determine whether a more complete relationship between cell size and receptor expression exists. These receptors may be masked by the glycoprotein LHY-peabody-Q which is cleaved by the enzyme shermanase. If this is the case, then receptor expression may be independent of cell size and the experiment may only be showing a masking effect of the glycoprotein. What you are asking is, 'Is there a proportional (linear) increase in Boris-X1 receptor numbers with increasing concentrations of shermanase?' What the experiment is asking is, 'Overall, is there a linear relationship between shermanase concentration and fluorescence in the moose-squirrel antibody assay of these amoebae?'

Statistically, you are trying to determine the chances of observing a linear relationship with a positive slope as steep as that observed, given that no such relationship actually exists, i.e. receptor expression and shermanase concentration are actually unrelated.

Assumptions

Before considering the statistical details of linear regression, its assumptions, calculations, etc. we need to consider a more fundamental scientific assumption that underlies the basic question. By proposing that a relationship exists between two variables, **and by assuming a shape to that relationship**, such as a line, you are implicitly assuming that a phenomenological dynamic underlies the process represented by the data. In the introduction to this Chapter, we alluded to the mathematical relationship one would expect to observe between height and weight. The mathematical expression of that relationship was linear, meaning that weight showed a proportional dependence upon height (the dependent and independent variables required before anyone can draw 'cause and effect' conclusions from a data set). Mathematically:

$$\text{Wt} = (\text{proportionality constant}) \bullet \text{Ht} + \text{baseline}$$

This equation states explicitly that for every additional unit of height, a subject will also show an additional given number of units of weight. The measures have units, say inches and pounds, and the proportionality constant is therefore given in pounds per inch. This underlying assumption carries with it biological implications which must be accounted for. Recall, that we were able to argue that although weight is actually proportional to volume, because the radius of the average human is fairly constant, height is the primary factor in determining weight gain. Extrapolating this to the hypothetical experiment described above, you must now argue that for each unit increase in shermanase concentration there is a proportional increase in proteolytic activity, resulting in greater antibody binding and thus, greater fluorescence. The proportionality constant (i.e. the slope of the regression line) estimated from these data is therefore going to be measured in fluorescence units per micromole of shermanase. Implicitly, that means that the assay design must ensure that there are no steric hindrance issues for the shermanase, that the levels measured are not below the quantifiable limits of the assay, and/or that concentrations are not at or above the saturable limits of the amoeba population.

The statistical assumptions underlying linear regression are:

(1) the data are sampled at pre-specified levels of the independent variable;
(2) the data are normally distributed at each sample level;
(3) the variances of each underlying population at each level are equal (homoscedasticity);
(4) the relationship between the independent and dependent variables is linear.

You should see that these assumptions and those outlined above for the one-way ANOVA are almost identical. The latter assumed that the sample data obtained from all independent treatment groups drawn from underlying populations which were both normal and homoscedastic. Assuming the null hypothesis for the inferential test, that all the means are equal, was equivalent to saying that if each level of the independent variable (in this case concentrations of shermanase) was considered an independent treatment group, then a line passing through the samples would have a true slope equal to zero.

Calculation of the slope and intercept

While a line is usually described by mathematicians in the form $Y = mX + b$, statisticians use the expression $Y = aX + b$ to model their sample data. The reason is that the equation tells us that Y, a random variable, is a linear function of X, a non-random variable (one with predefined levels and no variance). In other words, **populations** of responses are proportional to the levels of the independent variable. Solving this equation to give the slope (a) and the intercept (b) in terms of Y would show that they must also be random variables. The parameter estimates a and b, which are derived from the sample data, are therefore really just snapshots from two underlying populations, each with its own true mean, α and β. If we had used m to designate the slope, then the mean of the underlying slope population would have to be designated μ, the character already used to indicate the underlying mean of a sample population for all the inferential tools. Strange but true!

The method by which the best estimates for α and β are derived from a particular sample is called *least squares regression*. In essence, this method looks for a line that, when placed through the values of Y for all the samples, has the smallest sum of squared deviations about it. Conceptually the process proceeds as follows. A line is drawn through all the sample points. The line has a slope and intercept which you use to predict a value of Y for each level of X; for X^*, $Y^* = aX^* + b$. For each X^*, subtract each value of Y from Y^* and square the difference. Then add all these together to obtain a *sum of squares*. The 'best' line is the one that minimizes that sum.

This process ensures that we are able to minimize the *conglomerate* variance about the line. But why take the sum of squared deviations from the predicted values of Y^* – why not just sum all the sample variances at each level? The answer is that the implicit assumption about the data relationship, i.e. that it is a line, requires these data to follow a proportionality model of association defined by the slope of the line. That assumption is discussed further below.

The equation for calculating the sum of squared deviations is simply

$$\text{Sum of squares} = \Sigma(Y_i - Y^*)^2 = \Sigma(Y_i - (b + aX^*))^2$$

where each value of Y^* is measured at the appropriate X^*, and the sums of the deviations are made for each at its appropriate level. This is just a mathematical expression for what we said in English above. Performing the calculations allows the following expressions for the slope and intercept to be derived:

$$a = \frac{\Sigma(X_i - \overline{X})(Y_i - \overline{Y})}{\Sigma(X_i - \overline{X})^2}$$

$$b = \overline{Y} - a\overline{X}$$

where $\overline{X}$, $\overline{Y}$ are the grand means of the two populations taken separately.

The numerator of the expression defining the slope, a, looks very much like the numerator of the equation defining r, the sample correlation coefficient. This should come as no great surprise since it assumes the same kind of data relationship i.e. linearity and normality of each Y. In fact, we will be using a correlation like r^2 as a measure of how well the line fits our data. Accepting that the mathematics is beyond the scope of this book, the regression line must pass through the point $\overline{X}$, $\overline{Y}$. What this procedure is doing is fixing a line at the point $\overline{X}$, $\overline{Y}$ and spinning it about this center like a propeller until the sum of squares is minimized. When we have minimized that conglomerate variance, any noise remaining must represent that inherent in the Y populations. This noise provides a pooled estimate of the underlying variance in Y that should be observed at each level of the test. This is why we assumed that the variances of Y measured across all levels of X are equal. This process is shown in Figure 26.

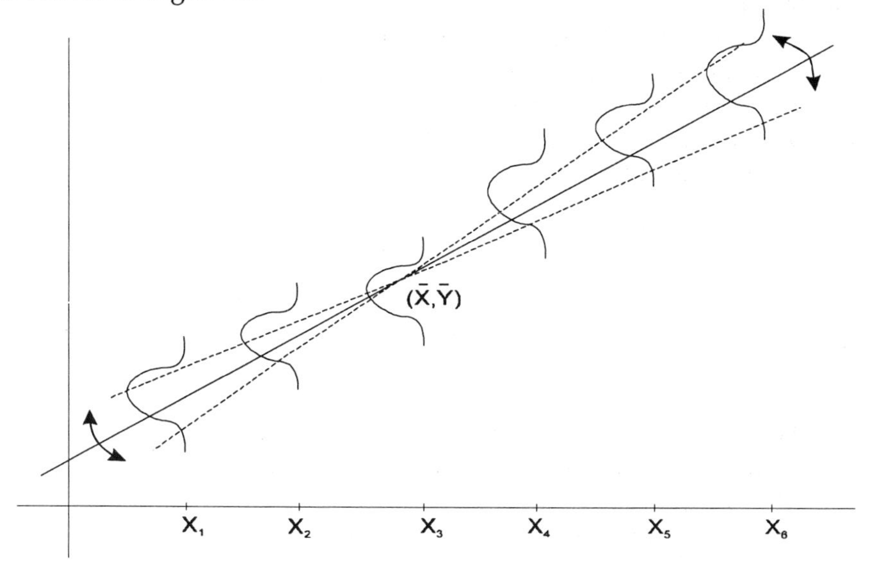

Figure 26

Since a and b are random variables and are only estimates of α and β we must account for the uncertainty about Y with confidence regions about a and b. The *standard error of the estimate* is the term used to quantify that uncertainty. It is related to all the other standard errors we have derived for the other estimates, μ, π, and ρ. The error estimates for

a and b should, therefore, explicitly depend upon the best guess of the underlying noise in the Y population. That estimate is given by a new variable, $s_{y.x}$, *the standard deviation of the regression,* which is calculated as follows:

$$s_{y.x} = \sqrt{\frac{\Sigma(Y_i - Y^*)^2}{(n-2)}}$$

$$= \sqrt{\frac{\Sigma(Y_i - \overline{Y})^2 - a\,\Sigma(X_i - \overline{X})(Y_i - \overline{Y})}{n-2}}$$

$s_{y.x}$ is conceptually similar to the pooled error estimate used to test for main effects in the one way ANOVA. By assuming homoscedasticity and treating each level of X as a different treatment arm the ANOVA can again be linked to the linear regression.

The first equation is clearly more intuitive than the second: it allows an estimate of the pooled variance to be derived from all the squared deviations. In reality the second equation is more computationally efficient, and is the one used to perform the relevant calculations in most software packages.

Once $s_{y.x}$ is known, the equations for the standard errors about a and b are actually quite simple:

$$\mathrm{SE}(a) = \frac{s_{y.x}}{\sqrt{\Sigma(X_i - \overline{X})^2}}$$

$$\mathrm{SE}(b) = s_{y.x}\sqrt{\frac{1}{n} + \frac{\overline{X}^2}{\Sigma(X_i - X)^2}}$$

One more piece of information is needed to complete the analysis of these data. Have you asked why the sample correlation coefficient, r, is not used to determine how well these data track each other? After all, the ability to interpret r correctly depends only upon assumptions of bivariate normality and randomness in X. We have, in fact, implicitly used r to calculate $s_{y.x}$ by including a cross-product term in its calculation. Therefore, intuition and logic tell us that this added bit of information could be quite valuable in evaluating our new regression results.

The calculation of r^2 for the regression is identical to that given for the correlation coefficient above. In this case, though, it is a measure of the *goodness of fit*, and can be used to decipher how much noise is due to the values of Y as they track those of X (after all it is implicitly assumed that $Y = aX + b$). The values of Y at each level of X are random variables drawn from underlying populations with their own means and variances. Assuming a positive slope, the means of Y should vary proportionately as the values of X increase. The variances about these means are assumed to be equal, but the conglomerate variance increases as the means move. [Hint: think about projecting all these moving normal distributions back onto the y-axis in Figure 26. The conglomerate looks fairly noisy!] What r^2 indicates is how much of that conglomerate noise is due to the drifting of the means, i.e. how much is due to the purported linear relationship, and how much is just noise.

The tests – the null hypothesis and its interpretation

We now have an estimate for the true slope, α, in the random variable a, and the true intercept β, in the random variable b, and we have a standard error about each.

When testing the slope, the most common null hypothesis is given as:

$$H_0 : \alpha = 0$$

This says that the slope of the line defining the linear relationship between the independent variable (in this case, levels of shermanase) and the dependent variable (the observed fluorescence, a random variable) is zero. In terms of the one-way ANOVA, what we are saying is that the means of observed values of Y sampled at each level of X are equal. In scientific terms, we are saying that there is no relationship between X and Y. What our test is asking is 'What are the chances of seeing a slope as steep as that observed, **given that there is no proportionality (linear) relationship between the independent and dependent variables?'** The test is given by the following equation:

$$t_{n-2} = \frac{a}{s_{y.x} \,/\, \sqrt{\Sigma(X_i - \overline{X})^2}}$$

The test is much like all the others outlined above. It assumes that there is a population of slopes, i.e. there is a random variable called 'slope', and that the mean of that population, α, is zero. The standard deviation of the population is represented by a function of the pooled sample standard deviation, $s_{y.x}$. The test is just asking how likely it is that a single realization a, drawn at random from this kind of population, would be as large as the one actually drawn.

Since our null hypothesis assumes that $\alpha = 0$, a is the distance of our estimated slope from α, i.e. our distance from zero. The units for slope in this case are just fluorescence units per micromole shermanase. The slope distance still has to be normalized by the noise metric. This normalization, similar to those used for the other inferential tests, is just the ratio given on the right hand side of the equation. The normalized distance is the number of noise units our estimated slope is from zero. Now we need to know the likelihood of being 'this far away' from zero. Linking the normalized distance to a probability is accomplished using the t-distribution with n–2 degrees of freedom (n–2 because two parameters were estimated from the data: the slope and the intercept).

If the t-value is large enough, the likelihood of a slope being as large as that observed, given that it was drawn from a population with a mean of zero, is quite low, say < 0.05. In that case the null hypothesis will be wrongly rejected, less than once in 20 times.

The most common null hypothesis for testing the intercept is:

$$H_0\colon \beta = 0$$

This says that the linear relationship between the independent variable (levels of shermanase) and the dependent variable (the observed fluorescence, a random variable) passes through the origin. In scientific terms, this says that there is no basal or background level to account for when defining the relationship between X and Y. What the test is asking is 'What are the chances of seeing an intercept this large, given that the true intercept is zero?' The test is given by the following equation:

$$t_{n-2} = \frac{b}{s_{y.x}\sqrt{\frac{1}{n} + \frac{\overline{X}^2}{\Sigma(X_i - \overline{X})^2}}}$$

This test can be interpreted in the same way as the test on the slope: what are the chances of drawing, at random, a value for the intercept as large as b from a population with a true mean of 0? Again we use a t-test with n–2 degrees of freedom. If the value of t is large enough, then the probability of seeing a realization from the population of all intercepts 'this large' would be low. In that case, you may be willing to reject the null hypothesis and be wrong a small percentage of the time.

Returning to our example

The results of the experiment outlined above yield the summary statistics shown in Table 31. Although we calculated the regression equations using a standard software package, we will present all the components needed to estimate the slope and intercept of this regression line. The value of the estimated intercept is 123.26 and the estimated value of the slope is 12.1. The value of $s_{y.x}$ is 13.74. The square root of the squared deviations from the grand mean of the concentrations, $\sqrt{\Sigma(X_i - \overline{X})^2}$, is 86.29. The denominator for the equation that tests the slope is thus given as: 13.74/86.29 = 0.158. 12.1 divided by that (12.1/0.158) = the calculated t-statistic, 75.994. The slope estimate of 123 is, therefore, about 76 noise units away from zero. The p-value associated with this t-statistic with 248 degrees of freedom is much less than 0.01. Similar calculations for the intercept yield a t-statistic of 93.70, which also yields a p-value much less than 0.01.

Table 31

	Concentration (μM)	Fluorescence (x100)
Mean	6.20	198.31
SD	5.47	67.57

These statistics tell you that the slope is positive, that there is a high level of basal fluorescence in the assay (of about 12 000 units), and that the likelihood of seeing data with slopes this steep and intercepts this high, given that the null hypotheses are true, is less than 1 in 100 for each. In fact, it is much less than 1 in 1000. The r^2 for this regression line is 0.96, suggesting that these data are highly linearly correlated: 96% of the conglomerate noise is due to the linear relationship between the fluorescence and the concentration of shermanase. The remaining 4% of the noise is due to the inherent variance in the underlying fluorescence populations.

Caveats and words of wisdom

The above discussion depends upon the implicit assumption that the data analyzed follow a linear relationship. What would we expect to see if they did not?

First consider the predictors, the values of Y^*. You would not really expect the values of Y for every sample at each level of X to be exactly equal to its appropriate Y^*. By definition, therefore, the sample contains a set of 'misses'. These misses, called *residuals*, are all the deviations from the predicted values of Y^* minimized in the sum of squares to obtain the best line possible. Theory tells us that because the values of Y were drawn from underlying populations that were normally distributed at each level of X, the misses about the line at each level should also be normally distributed. If they are not, then there may be a problem using a line to fit the data in the first place. *A residual analysis* is beyond the scope of this book, but this should be kept in mind when performing linear regressions.

Many sets of data show a relationship that is more complex than can be explained by a simple line. One of the many reasons for this is that the relationship between the variables is non-linear. Non-linear regression is discussed in the next section. Another explanation is that the dependent measure may depend upon more than one independent variable. This is a first look at an area called *multivariate statistics*, which is well beyond the scope of this book. However, there is one technique worth mentioning here which may prove useful when dealing with linear relationships between many variables.

Let us take tumor doubling time as an example, and assume that this depends upon on how fast individual tumor cells divide and how quickly they become quiescent. The doubling time would then be a function of both variables. Mathematically, this can be expressed as:

$$\text{Doubling time} = f(\text{birth rate, quiescence rate})$$
$$= aB - bQ + c$$

In this case, the shape of the function, f, is linear, but doubling time depends proportionally on both birth and quiescence rates. A technique called multiple regression has been derived to handle this type of problem. The details of the technique are beyond the scope of this book, but any good statistical consultant should be able to help you with this kind of analysis.

Finally, suppose you use a linear regression to analyze data from two random variables (such as height and weight in the example given in the last section), i.e. when, in fact, you should have been performing a correlation type analysis. While this may not be the end of the world, if you are going to use the wrong tool for the job, at least know what its limits are. Regression assumes a number of things that correlation does not. The interpretation obtained may be right, but it may not be as clean or efficient as if the right tool had been used.

COMPLEX DATA RELATIONSHIPS: MATHEMATICAL MODELS AND NON-LINEAR REGRESSION

Terms you should learn:

Parameters and parameter estimates
The law of parsimony
Model identification
Non-linear regression
ED_{50}, EC_{50}, IC_{50} and potency
Asymptotic standard error

Concepts you should master:

The mathematical model, first principles and philosophy of exploration
The sigmoidal curve in dose response and what it 'means'
The dose response and Michaelis–Menten dynamics
Non-linear regression and the minimization of the sum of squares
Parameter estimates, random variables and confidence intervals

Foreword – the mathematical model, first principles and philosophy of exploration

Up to now it has been assumed that the response variable (fluorescence) and the independent variable (concentration of shermanase) vary proportionally (i.e. they are linearly related). That implicit assumption provided the entire theoretical framework for linear regression as an estimation procedure, and the inferential tests that followed. If the relationship did not follow that simple proportionality constraint, performing a simple linear regression may have resulted in estimates for the slope and intercept that were quite misleading. Consider, for example, the following obvious case. The curve shown in Figure 27 was generated in an immunoprecipitation assay which measured precipitate vs. antibody concentration. Clearly, the amount of precipitate generated in this assay is not proportional to the amount of antibody added to the mix. A simple line fitted to these data would have given an estimated slope of zero and an estimated intercept of about 60 units of precipitate, which is, of course, nonsense. Since the relationship between these two variables is clearly non-linear, what we call biphasic, what do you think it should be? This leads us into a discussion of philosophy which

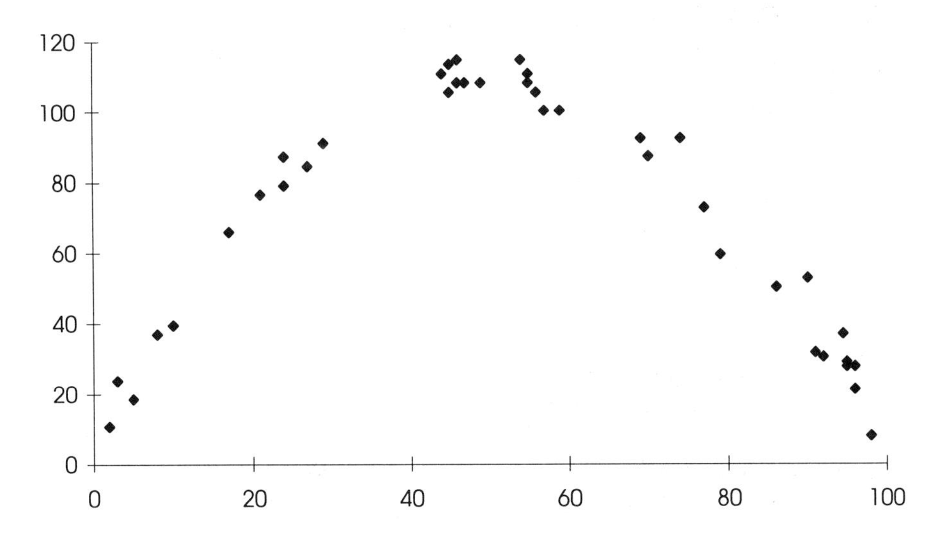

Figure 27 Precipitation vs. antibody concentration

basically crystallizes into the chicken-vs.-egg conundrum. In order to make scientific assumptions about the relationship between an independent and a dependent variable, is it not necessary to know, *a priori*, what that relationship should be? Is there not some mathematical formula, somewhere, that could be used as a paradigm for how Y varies with X?

Certainly, in a utopian world, there would be a catalog or library of such mathematical models to choose from so that fitting data would be just a matter of matching particular variables to a 'reasonably' shaped function. But what if there is more than one 'reasonable' shape from which to choose? Do all the models 'mean' the same thing? – i.e. do the *parameters* peculiar to each model all carry the same clear biological meaning that you want to communicate? (By parameters we mean constants in the model which modify its shape, like the slope and intercept of a line. You will see more examples of parameters when we discuss the dose response below).

Statistical estimation procedures are investigative tools as much as they are analysis devices. They can be used to explore the relationships inherent in data so that the 'correct' model can be chosen when you have no idea where else to begin. This exploratory role must, however, be tempered by good scientific judgement, and the model eventually selected should be one that can be defined, defended and interpreted intelligently. One rule of thumb is to use the most conservative model that explains the data adequately. This rule, known to mathematicians as *the law of parsimony,* says that the model used should make the fewest assumptions necessary, but no fewer. This field of statistics is known as *model identification:* the intricacies of this process are beyond the scope of this book.

Once the 'correct model' has been defined, whether by *a priori* logic and derivation or via model identification procedures, it is still necessary to estimate the model parameters from the data. This procedure is called *non-linear regression.*

An example: the dose response

One of the most common relationships explored in biology is that between the concentration or dose of an active drug or bioreactive molecule and the response it elicits. Typically, the response profile is an S-shaped or *sigmoidal curve.* The curve is non-linear and is defined by four parameters, which will be discussed below. First, look at an example (Figure 28).

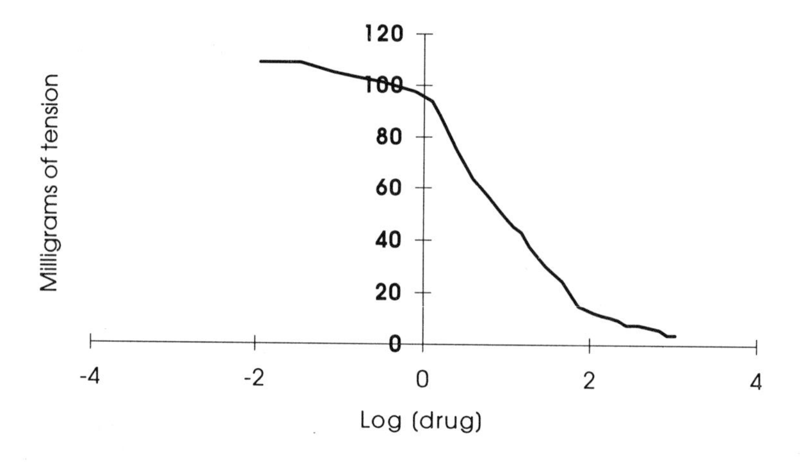

Figure 28 Sigmoid curve

Without indulging in mathematical minutiae, what does a curve like this tell you? First, it goes from a plateau on the extreme left hand side to a shelf on the extreme right. These dynamics suggest that the effect observed is saturable, i.e. no matter how much more drug is given, after a certain point a region of diminishing returns is reached. Mathematically, this saturable kind of behavior is called *asymptotic,* meaning that, in theory, the shelf can be approached as closely as you want, but it will only be reached when there are infinite levels of drug in the assay. We know that is not true – eventually the ability to detect these small diminishing differences will be exceeded. The model is an idealization of a perfect mathematical relationship. But the **form of the model** which must be developed must, in theory, be able to handle these kinds of dynamics in order to obtain a relationship that is both mathematically conservative, i.e. obeys the law of parsimony, and scientifically accurate.

The center part of the curve is linear (at least in so far as we are measuring effect vs. the log of the concentration). This indicates the existence of a proportionality between the dependent and independent variables for a significant portion of the data. That is very important both statistically and scientifically. The shermanase experiment was designed to exclude concentrations of shermanase either above or below the quantifiable/saturable limits of our targets. The linear regression was sufficient only in so far as there was a reasonable goodness of fit to our line. Problems occur when you extrapolate beyond the range of the data.

If you fitted a line only to the linear part of a saturable function and tried to project above the upper plateau to 'estimate the intercept' this estimate, which depends explicitly on the linear relationship underlying

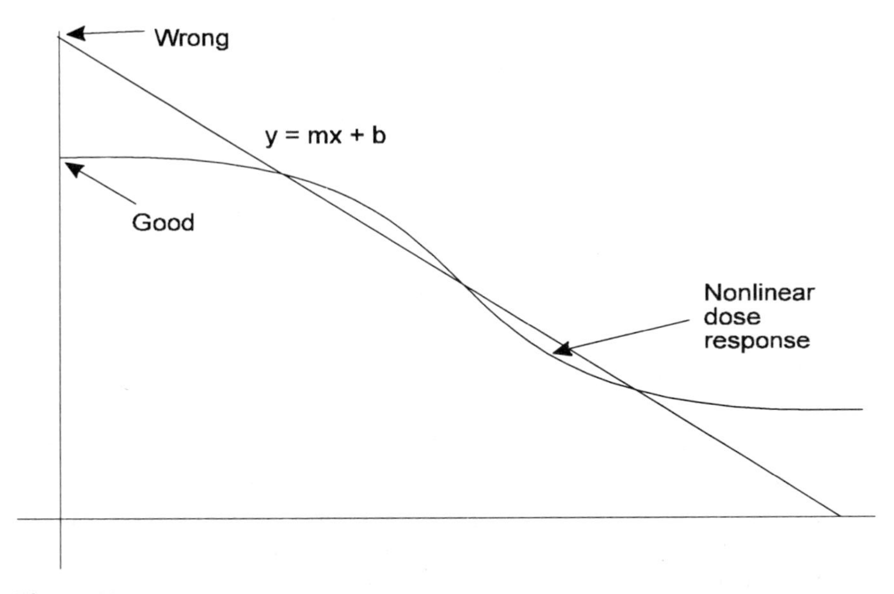

Figure 29

the regression, would overestimate the level of the upper plateau and completely ignore the saturable nature of the dynamics involved (see Figure 29).

A model of these kinds of dynamics can be derived from first principals and it is both mathematically and scientifically valid. The model is completely described by four parameters, each of which carries with it a well-defined biological meaning.

The Michaelis–Menten equation and the dose response

The model which will be derived is based upon the familiar mass-action equations of Michaelis and Menten which describe the dynamics of an ideal enzyme. Their equation is:

$$V = \frac{V_{\text{Max}}\ S^N}{K_{\text{M}}{}^N + S^N}$$

This says that the velocity of product production in an enzyme system, V, depends upon the amount of substrate in the assay (S, the dependent variable), and can go no faster than V_{Max}, the maximum velocity of the enzyme reaction (the first parameter). V_{Max} is a scientific measure based completely on the laws of thermodynamics. Dividing the numerator and denominator by S^N shows that the point at which $S = K_{\text{M}}$ is half way up the curve to V_{Max}. This is the second parameter of the model and is known as the Michaelis–Menten constant. It acts as a locator on the S-axis and indicates the affinity of the enzyme for the substrate: as the value of K_{M} increases more substrate is needed to move further along the response curve. N is a curvature parameter: the greater its value the steeper the response curve.

How does this relate to the dose–response function we were looking at earlier? In fact, both mathematically and scientifically there is a great similarity between these two dynamics. If you can believe, just for a moment, that the same physical/chemical laws of thermodynamics that dictate binding in the enzyme system dictate binding, signalling, etc. at the target site in or on cells, then the activity measured should have a shape similar to that of a Michaelis–Menten curve. The difference is that the inhibition of cellular activity, secretion rates of a hormone, etc. are being measured rather than the production rate of an enzyme product. Therefore, rather than estimating a parameter like V_{Max} you would estimate one which yields the maximum activity observable in the cellular system. The model could then be rewritten as:

$$\text{response} = \frac{\text{max activity}}{\left[\frac{\text{ED}_{50}}{\text{dose}}\right]^N + 1}$$

The new parameter in this model is called the ED_{50}, and is the dose corresponding to the point half way (50%) along the response curve. Conceptually, the ED_{50} acts the same as the K_M parameter in the Michaelis–Menten equation, locating the curve on the dose axis and providing us with a measure of *drug potency*.

What if there is a basal level of activity inherent in your assay system – as is usually the case when experiments are performed in cells, tissues or whole animals. That basal level is accounted for by adding another parameter to the model as follows:

$$\text{response} = \text{basal activity} + \frac{\text{max activity}}{\left[\frac{\text{ED}_{50}}{\text{dose}}\right]^N + 1}$$

Now look what happens as we vary dose. If N is positive, as dose approaches zero, ED_{50}/dose approaches infinity, the fractional term moves towards zero, and response reaches its basal level. As dose approaches infinity, ED_{50}/dose moves towards zero, and the fractional term approaches the maximum, and the sum becomes Max + basal. If N is less than zero the dynamic is reversed.

There is nothing magical about using dose and ED_{50} as the dependent variable and measure of potency: they are locators, just like K_M. If a concentration in vitro is being studied rather than a dose in vivo, the dependent variable is denoted 'concentration', and the potency parameter may be either EC_{50} or IC_{50} depending on whether you are looking for effective responses (EC_{50}) or inhibitory responses (IC_{50}). The point is that the same equation fits all.

Statistical issues

Statistically, the regression analysis is quite complex and it cannot, in fact, be solved with simple equations for each of the parameters (EC_{50}, maximum activity, basal activity, and N) like those that we used to obtain the slope and intercept of the line. A computer is needed to fit a non-linear equation to a data set. The idea of fitting the data is identical whether the shape of your response function is a line or a curve. What the computer program is doing is looking, iteratively, for the best point in a four-dimensional space, i.e. the best quadruplet (EC_{50}, maximum activity, basal activity, and N), so that the sum of the squared residuals is minimized. This is exactly the procedure employed when we took a line through the point $(\overline{X}, \overline{Y})$ and spun it like a propeller until the sum of squares was minimized in the two-dimensional (linear) space. By mimicking this procedure, we again minimize the conglomerate variance in the system, and are able to obtain a pooled estimate for the underlying variance in the random variable, 'response'.

The computer programs which perform non-linear regressions are quite complex, and require a certain amount of expertise in their programming and interpretation: this is well beyond the scope of this book. Working closely with a statistician, however, you should be able to derive estimates for your parameters based upon reasonable mathematical models of relationships obtained from your data.

There is one more point to be addressed. The parameters fitted to the 'perfect model' are, of course, only random snapshots drawn from four underlying distributions based on the four random variables 'potency', 'maximum', 'basal', and 'curvature'. They carry with them, therefore, as much uncertainty as is inherent in the underlying populations. The error estimates for the parameters of a non-linear regression are termed *asymptotic standard errors (ASE)*, and can be considered the estimation equivalent to the regular standard error of the mean calculated for the descriptive statistics above. The 95% confidence intervals about each parameter, say the EC_{50} in this case, can be given by:

$$[EC_{50} - 1.96(ASE(EC_{50})),\ EC_{50} + 1.96(ASE(EC_{50}))]$$

The factor of 1.96 comes from the fact that 95% of the area of the standard normal z-distribution is accounted for by 1.96 standard error units to the left (minus sign) and right (plus sign) of the center. A slightly more conservative confidence interval would be derived by using the t-distribution rather than the z-distribution. In this case the multiplying factor would be larger, its magnitude depending upon the degrees of freedom. In this instance the number of degrees of freedom is just equal to the number of data points observed minus four, paying one degree of freedom for each of the four parameters being estimated.

4 DESIGN OF A STATISTICAL EXPERIMENT

Terms you should learn:

Research hypothesis
Study population
Confounding
Specificity, accuracy and precision
Experimental strata and covariables
Sequential design
Completely randomized (one-way) and block design
Latin square design
Multifactor designs: factorial and fractional factorial designs

Concepts you should master:

The study population and its underlying distribution
Bias and experimental conduct
Experimentation vs. characterization
Elements of measurement
Choice of controls
Statistical power, margin of error, Type I and Type II error, and the study plan

The statistical experiment

A scientific experiment is a procedure used to answer a question by generating and analyzing data. It is a physical act, depending upon the trained hands and mind of the scientist to make it happen. The statistical experiment, on the other hand, is a process made up of several stages:

(1) formulation of the experimental goal;
(2) development of a statistically valid experimental design;
(3) careful conduct of the experiment;
(4) a thorough analysis of the data generated by the experiment.

Executing all these steps will allow you to answer a scientific question as accurately as possible. The product of this process is information, and like any good process, the quality of the final product depends upon the care with which each stage is executed . This section discusses, in detail, each stage of the process, demonstrating how careful conduct of the statistical experiment will help the actual experiment and realize the final goal, the gathering of quality information.

Step 1: the experimental goal

The experimental goal is sometimes called the *research hypothesis,* and is a rigorous statement of fact about some characteristic of the underlying population being studied. To understand it fully requires the definition of two very important terms. The first, known in statistics as the *study population,* is the universe of measurements made on objects being studied. Statisticians also call this the underlying target population. It is associated with an underlying distribution. The study population is, therefore, not batches of a compound or groups of animals, but the measurements made on these experimental units. Recall that we based our entire discussion of the inferential paradigm on this population. Results obtained from the sample population used in the experiment must eventually be generalized to this population. The second definition is that of *the characteristic of the study population.* This is any attribute of the **distribution** of these measurements which is scientifically interesting. You may want to study the average (mean) yield of a fermentation process, or perhaps you are interested in the variability of that distribution. Sometimes you may wish to know both – estimates of both are needed before the inferential tools described earlier can be used. Knowing the characteristic(s) to be addressed in the statistical experiment will help in the selection of the appropriate tools for collecting measurements and in the selection of an ideal statistical experimental design.

The following example will be used in the discussion of each stage of the process. An investigator wishes to show that an anti-hypertensive drug is effective in dogs. To initiate her experiment, she has to define what 'effective' is: the compound must reduce blood pressure in dogs by at least 20 mmHg over that seen in placebo-treated animals. In this example, the characteristic of interest (in statistics, *the population parameter*) is the average difference in blood pressure reduction between drug and placebo-treated animals. The explicit goal is to demonstrate an improvement in drug-treated animals by as much as 20 mmHg over controls.

Step 2: the statistical (experimental) design

A *statistical (experimental) design* is a plan. It includes, in clear and concise language, the details of how animals will be assigned to treatment groups (e.g. the randomization scheme) and the conditions to be applied to all treatment groups to ensure they are treated comparably (what we call applying the experimental conditions to the subjects of the experiment). Test subjects are called *experimental units* by the statistician. The experimental design must also include a sample size calculation to ensure that the experiment will provide enough satisfactory information to ensure that any conclusions drawn are probably accurate. We discuss the power calculation in detail below.

There are many ways in which to implement an experiment. For example, the effectiveness of a drug for reducing blood pressure could be studied in a number of different ways. You could study animals receiving only the experimental compound, or you could include a control group. Measurements might be made of post-treatment blood pressure alone, or you could look at changes from a baseline. Animals may even be treated once with the control compound and again with the test compound. The goal is to obtain the best information for the least investment of effort. The experimental design selected depends upon the experimental goal and the nature of the population being studied. It should be chosen on both the basis of practical considerations and the quality of information it can produce.

In our example, the investigator should probably consider an experiment design which ameliorates the high degree of variability in the distribution of blood pressure measurements between individual animals. She might plan to collect baseline and post-treatment blood pressures, calculating a change from baseline within each animal. Another strategy might be to use the same animals to study both the test and control compounds. The treatment effect must be large enough to be distinguishable from its own noise. Remember that the variance of the distribution of sample means decreases with increasing sample sizes – the same thing is true here. Sizes of treatment groups must be considered carefully: an experiment in which samples are too small can miss a practically meaningful outcome, while an oversized experiment is inefficient and may yield spuriously significant results, i.e. results which produce a statistically significant *p*-value but which are scientifically irrelevant. Remember, the *p*-value is a conditional probability, not a measure of scientific effect!

After considering all these elements, the investigator decided to measure blood pressure before treatment in each animal and at 30-minute intervals for 5 hours during infusion with either the drug or a vehicle control. She calculated that six animals would be needed to ensure 80% power of detecting a mean difference of at least 20 mmHg between treatment and vehicle control groups at the maximum blood pressure point ($\alpha = 0.05$).

This is a very complex statement of the experiment design and includes the criteria needed for power estimates and sample sizes. This will become clearer when we discuss the statistical power of an experiment, later. The point to make here is that the investigator has determined a plan (baseline measures, drug infusion, half-hourly sample intervals), and a sample size (six) which she believes will help to produce an answer to the original question.

Step 3: experimental conduct

The experimental plan is not complete until a decision has been made as to how the study will be administered, i.e. how the *experiment will be conducted*. Bias may be introduced from subtle influences of extraneous (environmental) factors or improper execution of the experimental plan, and biased data can easily flaw experimental results. To minimize the bias and balance environmental effects, the subjects (experimental units) should either be randomly assigned to treatment groups or drawn randomly as a representative sample from the study population. For example, animals receiving the same treatment that are all housed together are subject to hidden biases. Bias may also be a result of the way in which units are measured. This is especially true when measurement characteristics, such as run-to-run variability in an analytical test, influence the comparisons of interest.

In our example, the dogs are labeled 1–12 and assigned to receive either the drug or a vehicle control according to a random allocation schedule. Because this is a short-term study, the scientist is not too concerned about environmental factors such as housing. However, only six animals can be treated and monitored at a time. To eliminate measurement bias, she elects to make measurements from three animals from each group during each monitoring session. This will allay potential bias due to 'monitoring period' in the experiment: suppose the measuring devices (or animals) were heat sensitive, and all the drug-treated animals were tested in the cool of the morning while all the control animals were tested in the hot afternoon sun. There would be no way of telling whether effects observed were a result of drug effects or the heat. Such confusion of factors and elements is referred to as *confounding*.

Step 4: data analysis and conclusion

Data analysis is an accurate summary of the experimental data into some *information metric*. Examples of information metrics used throughout this book are the confidence interval and *p*-value – but what do we really mean by an information metric? What we are saying here is that the final span of the bridge between the underlying target population and our sample data needs to be completed. The process of data analysis allows random snapshots – data – to be put into a theoretical context for interpretation. The result of that interpretation is information. The information metric is the probability measure that indicates the likelihood of an interpretation being the correct one.

If the study has been well designed, and the experiment has been adequately administered, the analysis should follow seamlessly. In fact, a proper design should acknowledge the mathematical tools that will be used to analyze the data before the actual experiment is ever performed. Any deviation from the data analysis plan, like any other deviations in design, can hamper the investigator's ability to correctly interpret his or her results.

Some of the analysis tools available for summarizing and drawing conclusions from experimental data have been presented before, e.g. graphs and plots, *t*-tests, estimators and their confidence intervals. Linking them, a priori, to the design of the experiment, its conduct and its measurements means that the chances of using the wrong tool for analysis of the experimental results are much less than if the two processes are considered separately.

Back to our example

The goal of the experiment was to ascertain a difference in blood pressure reduction between drug-treated and vehicle-treated animals. Our investigator chose to represent her results with a 95% confidence interval on the mean difference in blood pressures. In other words, she wants to be 95% sure (her information metric) that the *true* mean of the differences in blood pressures (the population parameter) lies somewhere above 20 mmHg. The mean difference observed after analysis of the data gathered from the experiment was 28.7 mmHg. The confidence interval around this mean (22.5 mmHg, 34.8 mmHg), was derived using the standard error of the mean as described above. Since this interval excludes 20 mmHg, she felt confident that there was statistically significant evidence of a difference of more than 20 mmHg between drug treatment and vehicle control in dogs ($p < 0.05$). Careful application of each step in the process of a statistical experiment cedes a dependable marriage of practical and statistical significance in the performance of the actual experiment.

Design elements

Several design elements can affect the reliability of information obtained from a statistical experiment. Some of these, such as the sampling scheme used to collect the data and the sample size necessary to obtain reliable information, have already been mentioned in passing. Others, more subtle and frequently overlooked, are discussed here. These include the distinction between experimentation and characterization, elements of measurement, experimental strata and covariables, and choice of controls.

Experimentation versus characterization

One design element which relates to the scope of the conclusions that can be drawn from a statistical experiment is the difference between conducting an experiment and characterizing a population. In an experiment, only the experimental factor(s) should be allowed to vary; all other factors must be held constant. During characterization, non-experimental factors should be as varied as possible to represent the entire underlying population. What does all that mean?

An experiment is usually performed in an attempt to explore the influence of one or more factors on a particular population characteristic, e.g. the effect of a drug (the factor) on the mean or mean difference in blood pressure (the characteristics). This factor might be isolated in an attempt to abate the noise induced by non-experimental factors and to improve the likelihood of observing treatment effects. Thus, an experiment aimed at studying the effects of several different constituents on product stability might utilize a single lot of the active stock, and testing might be performed in limited analytical runs. This would reduce the variability inherent in the analytical method. This does not mean that a 'one-factor-at-a-time' experiment should be performed when different experimental factors are involved: the rule of thumb is that only random (environmental) elements which can contribute noise to an experiment should be controlled. Experimental factors which might systematically influence the process, such as all the various constituents relating to product stability, should be studied simultaneously (see Factorial Experiments, below).

The goal of characterization, on the other hand, is to describe the underlying distribution of measurements which evolve under the influence of both systematic (i.e. treatment) and random (i.e. environmental) effects. If any of these components is overlooked in the planning of an experiment the final data may not be representative of the true effects sought. Interpretation is then limited to the scope of only those factors actually included in the experimental plan. Thus, to estimate (i.e. characterize) the stability of a product, several lots should be included in the study design. If only one lot is studied 'lot' is no longer a factor in the design, and any conclusions drawn are relevant only to the lot studied. Nothing can be said about the population of material produced by the process.

Not being able to generalize the results of an experiment to an entire population because of a failure in the design is a special case of *confounding*. The example we provided earlier is a case in point: had our scientist not balanced her blood pressure monitoring so that treatment groups were represented in each monitoring session, i.e. they were monitored under comparable circumstances, then she could not have drawn any conclusions about whether the drug actually lowered blood pressure. Equally acceptable alternatives would be that the heat of the day, diurnal rhythms, differences in laboratory light levels, or handling by the afternoon vs. morning staff yielded the same results. By balancing the design the random factors of environment, handling, etc. were controlled for, and the results were free of confounding.

Elements of measurement

The quality of information derived from an experiment is affected by the selection of measurement variable. The overall noise in a statistical experiment is a composite of random variability, systematic treatment effects and measurement variability. A measurement variable should be chosen on the basis of its *operating characteristics*: *specificity, accuracy and precision.*

While these terms are usually associated with the process of validating an analytical method, they also apply to other situations in which scientific measurements are made. *Specificity* is the capacity to measure expressly the characteristic of interest, while *accuracy* is the ability to obtain, on average, the true value you seek. *Precision* relates to the inherent variability of the measurement. A measure that is precise will give the same results from measurement to measurement, although it may not be accurate (e.g. a precise scale which is miscalibrated yields a set of measures all within 1% of each other but systematically wrong). A measure that is accurate may not be precise (e.g. the electrical activity of a nerve bundle, while yielding an accurate picture of ion channel activity, is, due to limits in our technology, so noisy that the measurements are not very precise). Measures that are specific may be neither accurate nor precise (e.g. the effect of a drug that acts directly, and only, on an ion channel may yield physicochemical measures that are still very noisy, and, given the limits of our technology, possibly inaccurate). The optimal experiment requires your measurements be as specific, accurate and precise as possible.

These effects are reflected in the statistical power of an experiment: an experiment aimed at exploring the toxic effects of a compound in vivo could use either survival or weight loss as its measurement end-points. Survival is quite specific, but the measurement 'time to death' is not very precise, not because of limitations in the technology (after all, dead is dead) but because survival is such an inherently complex biological process. Weight loss, on the other hand, is likely to be more precise, and the chance of inducing excess noise in the system is minimized. With less noise it takes far fewer animals (and time) to obtain the same amount of suitable information.

Experimental strata and covariables

Strata and *covariables* are elements in an experiment which are not the primary factors (e.g. treatments), but which may exert an effect on experimental endpoints. If, for example, male and female mice react differently to treatment with a specific compound, sex should be considered as a stratum in the experiment. *Experimental strata* are typically categorical variables such as sex, formulation, container, etc. while *covariables* are continuous (i.e. measurable) characteristics of your population such as age, weight, pH and baseline response.

Strata and covariables which might have an effect on experimental results should be identified in advance. Recognizing these elements reduces experimental variability, enhances the ability to represent accurately the underlying population and isolates important experimental interactions. For example, simply eliminating sex as a source of experimental variability by identifying 'sex' as a stratum (and using the correct tool in the analyses) will allow proper conclusions to be drawn about drug effects taking each sex into account. If an outcome measure such as drop in blood pressure is strongly associated with an inherent characteristic such as weight, identifying 'weight' as a covariable, and adjusting the analyses to use this added knowledge can reduce the underlying noise in the experiment. The appropriate analysis tool for this situation is called an *analysis of covariance*. The details are beyond the scope of this book, but in essence it combines the ANOVA and correlation tools to minimize the variability due to weight.

Care should be taken with strata and covariables when running the actual experiment. To guarantee balance and to limit confounding, experimental units from each stratum should be randomized separately, and covariables should be collected as data throughout the experiment. There is, after all, no point in identifying a stratum or covariate if it is not then used to minimize noise.

Control groups

Experimental controls are an essential element in a well designed scientific experiment. They serve to validate the experiment, to define the study range and act as the 'gold standard' for comparison purposes. Experimental conditions other than treatment, such as environment or time, can affect responses in an experiment. The controls form the relative basis of response when all external elements are accounted for. They act as validation indices in the experiment: if the control subjects are not behaving as expected there may be something wrong with the entire experimental procedure. Under certain circumstances, particularly in analytical procedures, multiple controls are used to achieve these objectives. An assay or reference standard is used to calibrate test compounds, while an assay control is used to monitor the operating characteristics of the assay. Vehicle controls are fundamental to the design of in vivo pharmacology and safety experiments. When developing a new manufacturing process, a batch of material manufactured under the standard process can be the control against which all other processes can be compared for yield or purity.

Additional design elements

Many other elements should be considered when planning a statistical experiment. Below we work through a simple *power calculation*, deriving a sample size for an estimated mean based upon a fixed margin of error. The principle of restricting the margin of error holds for all experimental plans and is the basis for all sample size calculations. The equations that have been derived to calculate these sample sizes intimately depend upon the statistical design of the experiment, and can be quite complex. You may need the help of a statistician to utilize them fully.

So far we have assumed that our experiments are meant to show a difference between two groups, in this case, canine blood pressure. While this scenario is familiar to most of us, many circumstances warrant a different approach. Some experiments seek to demonstrate the *equivalence* of effects across experimental factors. For example, suppose we wish to validate that a new production process is equivalent to our present one, or to establish in vivo that the distribution of an active ingredient is equivalent over several formulations of a drug. The data analyses required to 'prove' equivalence are different from those needed to establish a treatment difference. The design of these experiments is also special: a statistician should be consulted for help in the details of an equivalence study. They are not as easy as they appear.

Sometimes, the sample size required to undertake an acceptable statistical study is too large to be practical. This may be due to inherent noise in the population under study. A *sequential experimental plan* can be used incrementally to improve the knowledge of a population's characteristics (especially the underlying mean and standard deviation), and may allow a smaller overall experiment to be performed. For example, if an experimental plan calls for eight dogs per group to study changes in blood pressure, data might be collected initially on four dogs per group, and an interim analysis of the results performed. These interim studies can give a rough estimate of the true means and standard deviations in each group. If the statistical power is not sufficient to establish the desired effect, four additional animals in each group can be studied to complete the experiment. There is a price to pay for a sequential study plan, and such studies should only be undertaken following expert advice.

Statistical power versus practical significance and sample size

An experiment which has too few subjects may miss a *practically significant* outcome, i.e. a difference large enough to be scientifically relevant. This is called a *Type II error* and represents the 'consumer's risk'. The better known name for this kind of error is a false negative, i.e. concluding that there is no difference between groups, when in fact one exists. On the other hand, an experiment which is too large may generate a spuriously significant result. Statisticians call this a *Type I error* or the 'producer's risk'. This kind of error is a false positive: a true difference appears to be present when in fact there is not. When an experiment is well designed (including a complete power calculation), statistical significance can act as a surrogate for practical significance. Therefore, a good design will acknowledge, *a priori*, the inherent risks, the Type I and Type II error levels, the underlying variability of the experimental measurements and the practical goal of the study.

Consider the following example. Suppose we are interested in estimating average dissolution of a tablet formulation, and we would like our estimate to be within a specified margin of error. The relationship defining the terms which limit the margin of error on the estimate is:

$$\text{margin of error} = \text{confidence} \bullet \frac{\text{measurement error}}{\sqrt{N}} \leq \text{practical deviation}$$

Thus, if we fix the practical deviation, i.e., how close we want to be to the truth, the margin of error can be reduced by:

(1) decreasing the confidence in the estimate – an undesirable strategy since confidence relates to the chance of drawing the wrong conclusion;
(2) reducing the measurement error – something we cannot do since it is usually a static property of the population;
(3) increasing the sample size.

Solving this equation for the required sample size, N, gives:

$$N \geq \left(\frac{\text{confidence} \bullet \text{measurement error}}{\text{practical deviation}} \right)^2$$

In our example, had we wished to estimate dissolution to within ±2% (practical deviation) with 95% confidence (a value of about 2 for 'confidence' in our expression), and had we 'known' that our measurement error is equal to 3%, we would need at least 9 (= $(2{\cdot}3/2)^2$) sample points to meet our power criteria. The numbers derived for 'confidence' in this equation come from the *t*- or *z*-distributions, and are the same as those used to obtain the *p*-values. Had we elected to be 99% confident about our result, 'confidence' would have a value closer to 3.

Design digest

Some commonly employed experimental strategies are considered here. Their applications, advantages and disadvantages are summarized, along with a basic schema for each. The sample size for each design is selected for illustration purposes only.

The completely randomized (one-way) design

The completely randomized design is the most commonly employed statistical design in pharmaceutical development. This is sometimes called a one-way design because, in general, N experimental units are randomized to k treatments which are defined by a single experimental factor.

As an illustration, consider a safety experiment in rats divided into groups of six animals each, to receive a vehicle control, a low dose, medium dose and high dose of a drug. A computer generated enrolment (randomization) scheme is shown in Table 32.

Table 32

Control	Low	Medium	High
14	5	11	12
2	17	16	3
10	13	1	18
24	8	21	7
19	20	4	23
6	9	22	15

Animal 1 is assigned to medium dose group, animal 2 is assigned to the vehicle group, and so on. The advantage of the completely randomized design is its simplicity. Many popular spreadsheet programs provide online resources to analyze data from a one-way layout. If only two groups are needed, say a control and one test group, the two-sample *t*-test or its Mann–Whitney non-parametric one-way ANOVA or the Kruskal–Wallis non-parametric test can be employed. These tools were discussed in detail earlier. The disadvantage of this design is that it is costly, frequently requiring larger sample sizes than some alternative strategies.

Other examples of the completely randomized design include comparative studies of different compounds in vitro, comparison of compound formulations based upon different excipients and cell culture experiments comparing different culture media supplements. Key to this design is that only a single factor is under investigation (e.g. compound, excipient, supplement). If the effects of several factors are to be studied a factorial design should be employed.

Block designs

Block designs can be used to reduce the effects of factors in an experiment which create variability. Two such designs are a *complete block design* and a *Latin square design*.

The complete block design is used in pharmacology to study the effects of increasing doses of a compound. Suppose serial doses (treatments) of a compound are administered to each of several tissues (blocks), as shown in Table 33. Treatments are administered in the same order in each tissue.

Table 33

Tissue	0.1	0.3	1	3	10
1	1	2	3	4	5
2	1	2	3	4	5
3	1	2	3	4	5
4	1	2	3	4	5
5	1	2	3	4	5
6	1	2	3	4	5

The primary advantage of blocking is that it controls the variability between factors in the block (here tissue to tissue) variability. The block design is one possible solution when non-experimental factors contribute a large amount of the variability to an experiment. If an assay exhibits large inter-run variation, then samples from multiple treatment arms can be submitted to the assay in carefully selected groups. A case in point is the balancing act performed by our scientist when she tested three dogs from each treatment arm in the morning and three more from each group in the afternoon. Time of day became a blocking factor. Similarly, when different devices are used in the conduct of an experiment, such as fermentors or tablet presses, it may be wise to identify these devices as experimental blocks. In addition to reducing experimental variability, block designs enhance the ability to generalize experimental results to wider portions of the underlying population.

This type of design is called a two-way design (here two-way refers to the two factors in the design, treatments and blocks) and can be analyzed as readily as the one-way design using popular spreadsheet programs. If there are only two treatment levels, a paired *t*-test or Wilcoxon non-parametric test should be performed. If there are multiple treatment levels, a two-way ANOVA or Friedman's non-parametric test can be used.

A disadvantage of the complete block design is that it does not control the order in which the treatments are administered (*sequence effects* in the experiment). This situation can arise when a drug accumulates in the blood stream or induces dose-dependent permanent changes in the host. A strategy which achieves balance in the sequence of treatments applied to the blocks is a *Latin square design.* Suppose three treatments (A, B and C) are to be administered to three dogs. We could give all three dogs Treatment A first, then B, then C. That is the block design. If, however, treatment A induces a permanent change in the test subjects (e.g. kidney damage or impaired liver function), the effects of B and C may be masked completely. Alternatively, if each dog received a different treatment first, and if treatments were changed for the second treatment period, and then again for the third, a scheme like that shown in Table 34 would be obtained. This design is frequently employed in animal pharmacokinetics experiments, where small numbers of animals are used to study several formulations of a drug. As with all block designs, the Latin square design helps ameliorate the effects of inter-animal variability.

Table 34

Dog	A	B	C
1	1	2	3
2	2	3	1
3	3	1	2

Multifactor designs (factorial and fractional factorial designs)

The goal of many experiments – drug formulation studies, product control processes and assay performance experiments to name a few – is to study the effects of multiple factors on a particular endpoint. These are all processes which are subject to the influences of a number of constituents and the conditions under which they are implemented. *Screening designs* are used as investigative tools to help establish which factors exert the most influence on a particular response. *Optimization designs* use that information to find the optimum settings of the most important design factors. These multiple effects could be studied one factor at a time, but such a strategy is both ineffective and inefficient. It takes more time and resources, and furnishes less statistical power, than a *factorial design.* It can also miss important interactions among the experimental factors, which in a screening study can result in misidentification of key factors, and in an optimization experiment, yield substandard performance in the process.

Factorial designs are used to study the effects of multiple factors simultaneously. For example, suppose we want to establish which of three excipients, a lubricant, a binder and a filler, has the greatest effect on tablet hardness. As a screen, we might try two levels of each factor, high and low. A scheme frequently used to outline this kind of factorial design is to depict the low level of a factor with the code –1, and the high level using +1. Center points could also be included in the design, values somewhere between high and low, and coded using the symbol 0 (see Table 35). This does not mean that no excipient is given for that treatment combination. Rather it indicates there is a level of a particular excipient somewhere between high and low which is believed to be a reasonable level to test. This type of design can provide valuable information regarding the variability of your process.

Table 35

Run	Lubricant	Binder	Filler
1	–1	–1	–1
2	–1	–1	1
3	–1	1	–1
4	–1	1	1
5	1	–1	–1
6	1	–1	1
7	1	1	–1
8	1	1	1
9	0	0	0
10	0	0	0

A simple design using three factors at two levels each results in 2^3 (= 8) design points. The inclusion of additional 'center points' (coded 0 in Table 35) increases the size of the experiment. An alternative way to depict a simple factorial design is as a cube (see Figure 31). The vertices of the cube represent the experimental runs, while the size of each circle is proportional to the response observed at the associated factor settings.

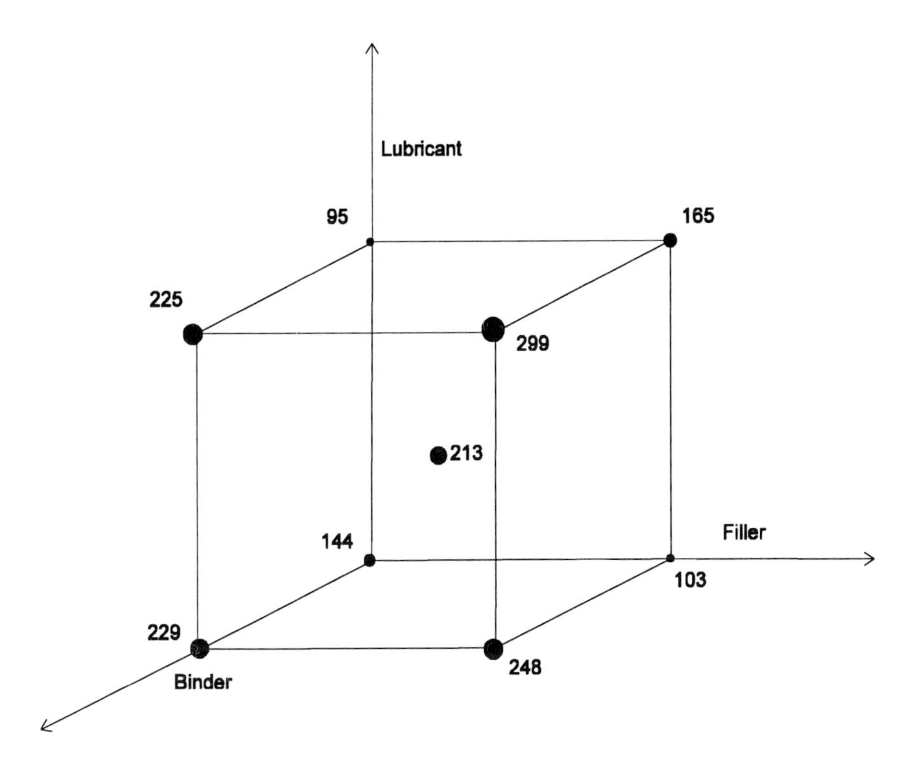

Figure 30 A cube representation of a factorial design

In general, you need 2^k runs plus center points to study k factors effectively. This can quickly become a fairly large experiment as the number of factors increases, and care should therefore be taken in choosing the experimental factors thought likely to exert the largest influence on outcome. One way to reduce the size of the experiment is not to limit the number of factors and levels, but to test only a select *fraction* of them from the overall factorial design. This is called a *fractional factorial design.* For example, a half-fraction of a 2^5-factorial design results in 16 experimental runs rather than 32. This reduction in the number of runs results, however, in the loss of some information. We cannot estimate all the possible interactions that exist between factors, i.e. they are *confounded* with each other. Fractional factorial designs and other statistically efficient strategies are extraordinarily complex and should only be used in conjunction with expert advice.

An optimization experiment is a different kind of exploratory procedure. To optimize a process effectively, factorial designs using more than two levels of each factor are usually employed. A 3^k-factorial design, known as a *central composite design,* is an effective choice. A detailed discussion of these designs is beyond the scope of this book, but any textbook on statistical experimental design will contain details.

Regression design

As described earlier, regression and correlation are the tools most commonly used to study the relationship between two variables. Simple linear regression and curve fitting are often used to fit models of kinetics over time or doses. Several design features need to be taken into consideration to employ regression effectively, such as regression scale which should be considered in selecting the number and spacing of doses and time points. Two commonly employed dose scales are arithmetic (linear) and geometric (sometimes called logarithmic). The choice of dose scale should yield as much information as possible, as economically as possible. In statistical language, we are trying to achieve parsimony between the regression points and the response variable. Arithmetic scaling is preferred when the response variable behaves arithmetically, i.e., linearly: dose 1 = dose 2 + X, dose 3 = dose 2 + X, etc. Geometric scaling should be used when the responses are multiplicative, i.e. dose 2 = twice dose 1, dose 3 = twice dose 2, etc. (see our description of the geometric mean for an example of this type of scale). The weights applied to arithmetically and geometrically scaled doses (where 'weight' is the importance of each measurement in the calculation of a regression line) are not equivalent (Table 36).

Table 36

Arithmetic		Geometric	
Dose	Weight	Dose	Weight
10	38%	10	31%
20	12%	20	19%
30	12%	40	3%
40	38%	80	47%

The arithmetic scale yields a more equitable distribution of weights among the regression points. Therefore, if a dosing regimen takes advantage of a dilution process, the geometric scaling will create an inherent imbalance in this distribution. A log transformation of the doses will then result in equal spacing among the regression points, and yield the same weighting as the arithmetic scale (see Table 36). The number of regression points selected should be sufficient to meet the objectives of the experiment. If the goal of the experiment is to measure the change in a response, two points is ideal, i.e. before and after. If more complex dynamics are suspected, more points should be utilized. Complex non-linear equations require special treatment. Regression points should be abundant in regions of rapid change, but can be less frequent where there is little variation in response. In correlation analyses, a broad range of responses will yield a stronger estimate of the degree of association between variables than those that are more closely spaced.

Repeated measurements

The more samples you take on a subject, the more you know **about that subject**. Because we are constrained to live in a temporal universe, samples represent dynamic responses in an assay system. Multiple measurement designs, wherein data are collected on the same experimental units, can therefore be valuable tools in safety, pharmacology, stability and other kinetic studies. We have already seen special cases of this design in the cross-over and Latin square designs presented above. Dose escalating studies represent another variation on the theme. There is also another reason for measuring the same subject repeatedly: in some instances between-subject variability can be lessened when multiple measures are made within subjects.

Repeated measurement experiments are one of the most statistically complex designs employed in research and development. Care should be taken when collecting and analyzing data from these types of studies. Repeated measurements on the same experimental unit are not independent, and this complicates the statistical analysis. Naive analyses, those conducted on data at each time point without correcting for the multiple comparisons problem, can yield spurious significances due to the large number of tests performed. Missing values can lead to information bias (see Figure 31), and a reasonable strategy to estimate these missing responses must be employed.

Sampling intervals should be selected both to be convenient and to provide the maximum information possible. Frequent observations should be collected over periods of expected rapid changes in the response, while fewer observations are needed when the response varies slowly.

Dose response

The dose response is one of the most useful tools in pharmaceutical research and development. A trend in response with increasing dose is an effective way to establish a minimum effective dose (MED) or a maximum tolerated dose (MTD). The dose–response relationship can be used in lieu of pairwise comparisons of each treatment group in a one-way design. A parsimonious summary of the effectiveness or safety of a compound can be obtained without the need for a dose-by-dose comparison with a control. Over the linear part of the dose range, a linear regression analysis may be enough to satisfactorily characterize the dose response of a compound. The regression can be further applied to obtain a powerful comparison of doses with a vehicle control. For example, consider a study of the effect of a drug on bleeding time in dogs. The doses chosen are a vehicle control (dose 0), 1 mg/kg, 3 mg/kg, and 10 mg/kg of an anit-coagulent. You are fairly convinced that these three doses represent the linear part of the curve.

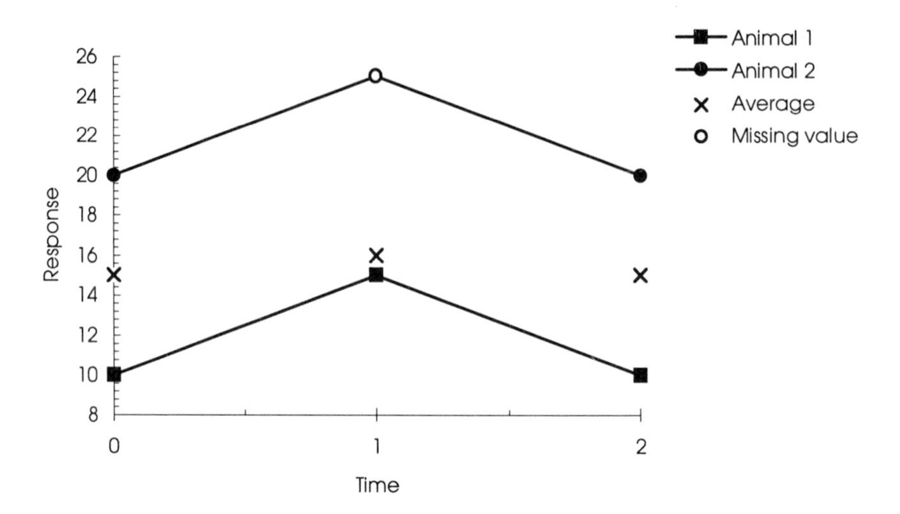

Figure 31

The bleeding times are equal to 22, 18, 32, and 56 minutes respectively. While the average bleeding time for only the high dose (10 mg/kg) is clearly statistically greater than the average response to the vehicle control (we performed a Dunnett's test after completing a one-way ANOVA here), a regression analysis yields a significant trend through the middle dose (3 mg/kg) of the compound. In this context, the maximum tolerated dose is unequivocal, whereas pairwise comparisons in the one-way ANOVA can lead to ambiguity in the conclusions.

Summary

The reward of sound statistical planning is quality information. Key elements of the experimental process are a well formulated objective, an effective experimental design, proper conduct of the experiment and careful analysis of the experimental measurements. From these four elements you are able to derive appropriate results and reach clear and accurate conclusions. Proper implementation of each step of this process helps to ensure the unambiguous quality of scientific information.

We have outlined some of the most important statistical design considerations required for a sound scientific study. Design elements which affect the reliability of your experimental results include the scope of your experimental inference, the selection of a proper measurement variable, the impact of experimental strata and covariables, use of experimental controls and determination of a sample size.

Several statistical designs have been discussed. The most commonly employed design in scientific research is the completely randomized or one-way design. This forms the basis for comparisons among treatment groups when no other factors influence the experimental outcome. A block or two-way design is used to account for variability from non-treatment related factors in an experiment. Multifactor designs are used in studies in which several experimental factors are applied simultaneously. These designs should be used in lieu of 'one factor at a time' strategies. Regression analysis is frequently employed to study kinetics of a single variable over time or doses, and correlation analysis provides information concerning the association between two measurement variables. Either of these analyses is best served by a carefully contrived experimental strategy. And finally, the design tool customarily used to directly assess the effects of a candidate compound is the dose response.

Index